CREATING THE TECHNOPOLIS

CREATING THE TECHNOPOLIS

Linking Technology Commercialization and Economic Development

Edited by
RAYMOND W. SMILOR
GEORGE KOZMETSKY
DAVID V. GIBSON

BALLINGER PUBLISHING COMPANY
Cambridge, Massachusetts
A Subsidiary of Harper & Row, Publishers, Inc.

International Standard Book Number: 0-88730-261-0

Library of Congress Catalog Card Number: 87-30739

Printed in the United States of America

Library of Congress Cataloging-in-Publication Data

Creating the technopolis.

   Proceedings of a conference held at the University of Texas at Austin, March, 1987, sponsored by the RGK Foundation, and others.
   1. High technology industries—Location—Congresses. 2. Cities and towns—Growth—Congresses. 3. Industrial districts—Congresses. I. Smilor, Raymond W. II. Kozmetsky, George. III. Gibson, David V. IV. RGK Foundation.
HC79.H53C74   1988   338.6'042   87-30739
ISBN 0-88730-261-0

CONTENTS

LIST OF FIGURES

LIST OF TABLES

PREFACE

Two key assumptions about the twenty-first century are the drivers for this book. First, we are on the threshold of a great technological era. In the United States and across the world, technology is profoundly affecting the way people think and act. It is dramatically altering the shape and direction of society. In the process, the rapid increase in and diversity of new technologies are changing the nature of competition. How communities, regions, and nations anticipate and respond to this new competitive environment, based on emerging technologies, will largely determine the health and viability of their economies.

Second, the nature of economic development has fundamentally and permanently changed. New institutional alliances are altering the strategy and tactics of economic development and diversification. New relationships between the public and private sectors—especially among business, government, and academia—are having far-reaching consequences on the way that we think about and take action on economic development.

These two assumptions are captured in the term *technopolis*. *Techno* reflects the emphasis on technology; *polis* is the Greek word for city-state that reflects a balance between the public and private sectors. The modern technopolis is one that interactively links technology commercialization with the public and private sectors to spur economic development and promote technology diversification.

Linking technology and economic development in a new type of city-state is an emerging worldwide phenomenon. Three factors are especially important in the development of a technopolis and provide a way to measure the dynamics of a modern technology city-state: the achievement of scientific preeminence; the development and maintenance of new technologies for emerging industries; the attraction of major technology companies and the creation of home-grown technology companies. This was the consensus that emerged from the international conference on which this volume is based that was held at The University of Texas at Austin in March 1987.

Because the idea of the modern city-state, as embodied in the technopolis concept, is new, even experimental, little hard data on or clear evidence predicts the progress of the technopolis. Consequently, important questions need to be addressed.

What are the real impacts and implications of the technopolis? Can the technopolis truly become a city of wisdom and light or does quality of life inevitably suffer where technology is predominant? What are the trade-offs involved in promoting economic development through technology? What are the lessons to be learned from the different technopolis models? Are there themes that span nations, or is the concept region and culture specific?

This book assesses the dynamics among business, government, and academia in creating and sustaining the technopolis. It brings together leading authorities from Europe, North America, and Asia to compare and contrast mature, developing, and emerging technopoleis in the world.

The chapters provide practical and theoretical insights on the factors that either enhance or inhibit the development of a technopolis. In the process, the focus on important issues concerning the role of universities, the functions of local, state, and national governments, and the role of both large and small technology companies.

ACKNOWLEDGMENTS

Many people and organizations helped make this book possible. We hope that it reflects well their dedication to better understanding the relationships between technology and society, the evolving direction of economic development, and the emergence of new institutional relationships.

We wish to thank Everett M. Rogers, Walter H. Annenberg Professor at the Annenberg School of Communications at the University of Southern California. He played a key role in the conception, planning, and development of the conference that led to this book.

We are grateful to Ronya Kozmetsky, president of the RGK Foundation. She provided invaluable support and organizational know-how for the conference.

We greatly appreciated the support and encouragement of Dean Robert Witt and Associate Dean Robert Sullivan of the Graduate School of Business at The University of Texas at Austin, and Professor W.W. Cooper, chairman of the school's Department of Management Science and Information Systems.

Becky Jessee of the IC^2 Institute deserves special recognition. She typed the manuscript through many revisions with terrific efficiency and good spirit.

We wish to thank Linda Teague of the IC^2 Institute for her assistance with the conference and her preparation of many of the figures in this volume and Elaine Chamberlain of the IC^2 Institute for her help in the early stages of conference preparations. We are also grateful to Cynthia Smith and Diane McDougal of the RGK Foundation for their invaluable administrative support.

We are indebted to the sponsors of the conference: RGK Foundation; the IC^2 Institute, the College and Graduate School of Business, and its Department of Management Sciences and Information Systems, at The University of Texas at Austin; and the Annenberg School of Communications at the University of Southern California. Their ongoing commitment to research, education, and application really made the conference and this book possible.

Ballinger is a great press with which to work. We appreciate the interest and encouragement of Carol Franco, president of Ballinger Publishing Company, the important help of Marjorie Richman, our editor, in preparing the manuscript for publication, and the expert assistance of Barbara Roth, the developmental editor.

We are very grateful to the authors whose contributions make up this volume.

Finally, each of us wishes to thank his co-editors for making the work on the conference, the book, and Chapter 10 such an enjoyable and productive collaboration.

INTRODUCTION

David V. Gibson, George Kozmetsky,
Everett M. Rogers, and Raymond W. Smilor

The concept of the technopolis—the modern city-state linking technology and economic development—is relatively new. Sometimes referred to as a *technology center* or a *high tech corridor or triangle*, the technopolis appears to be an emerging worldwide phenomenon. Because the concept is new, even experimental in some areas, there are unanswered questions and important issues about the dynamics of the creation and maintenance of a technopolis that need to be addressed and better understood. This introduction highlights these questions and issues.

The two oldest U.S. technopoleis*—Route 128 and Silicon Valley—did not receive much public recognition until the 1970s (in fact, Silicon Valley was not even so-named until 1971). Only in recent years have books begun to appear about this technopolis, and *Silicon Valley* has almost become a synonym for high technology throughout the world. A number of other Silicon Valleys are now emerging in the United States, in other advanced industrialized nations in Europe and Japan, and in certain developing nations like the People's Republic of China. Many of these technopoleis have developed spontaneously, while others (especially those launched in recent years) have been planned.

---

* The plural form of the Greek word *polis* is *poleis*. Therefore, we have chosen to use the plural *technopoleis* rather than other possible plural forms such as *technopolises* or *technopoli*.

ROLE OF THE RESEARCH UNIVERSITY

One of the oldest, and perhaps most successful, of the planned technopoleis in the United States is Research Triangle in North Carolina. This technopolis is the product of strong, consistent policy support by the state government of North Carolina, exerted in part through state government assistance to one or more of the three research universities involved in the Research Triangle: Duke University at Durham, the University of North Carolina at Chapel Hill, and North Carolina State University at Raleigh. The governor of North Carolina launched Research Triangle in the late 1950s, but progress was slow for twenty years. So the development of technopolis can be a long-term proposition. Here the key quesiton is, Do federal, state, and local government policies encourage the research university to engage in high technology development?

The strategy of developing a technopolis through improving a local research university has been followed in the past decade both at various U.S. locations, such as The University of Texas at Austin, Arizona State University in the Phoenix area, the University of Utah at Salt Lake City, and Rensselaer Polytechnic Institute (RPI) in Troy, New York, and in other nations like Sweden and England. Silicon Valley could not have occurred without Stanford University, and Route 128 could not have happened without the Massachusetts Institute of Technology (MIT). However, Sophia Antipolis in the French Riviera, Tsukuba City in Japan, and the Shenzhen Special Enterprise Zone (SEZ) in the People's Republic of China are not particularly centered on a research university, although a number of important government research institutes are located in Tsukuba, and a new technical university has been started in the Shenzhen SEZ.

A research university is an institution of higher education that emphasizes conducting research and teaching postgraduate students how to conduct research. There are about fifty such research universities in the United States today. A natural symbiotic relationship exists between the research university and high technology industries like microelectronics and biotechnology, at least in the United States. In nations other than the United States, where most research is conducted in government research institutes or by private industries, rather than in universities, is the research university a key ingredient in the emergence of a technopolis?

UNIVERSITY POLICIES TOWARD INDUSTRY

Further, to what degree can the emergence of a technopolis be speeded up by a local research university that has a favorable policy toward collaboration with private high tech companies in conducting research and in technology transfer? In other words, does a technopolis happen more rapidly when a policy favors (or requires) that a research university assist high tech firms, than when this collaborative relationship is not planned or directed? For example, Cambridge University did not have a favorable policy toward such collaborative technology transfer with surrounding high tech firms, although university authorities decided not to *oppose* it. Today there are over 400 high tech firms around Cambridge. In the 1980s many U.S. universities (especially state-supported universities) and several in Europe have adopted policies favorable to university/industry relationships in high technology development. Do these new policies dealing with patenting research results, university research parks, faculty consulting with industry, and so forth make any difference? Here the question is, Do university policies favorable to collaborative relationships with private industry actually encourage high technology development?

The role of the research university in the emergence of a technopolis may change over time. For example, the role of The University of Texas at Austin and of Arizona State University in Phoenix suggests that the university is crucial in the early stages. But such mature technopoleis as Silicon Valley and Route 128 now seem to operate under their own steam with the importance of Stanford University and MIT, respectively, now being much less than in the early days of these technopoleis. So here the question is, Does the role of the research university become less important as a technopolis matures?

COLLABORATIVE UNIVERSITY/INDUSTRY RELATIONSHIPS

An important event in Austin's development as a technopolis was the launching of the Microelectronics and Computer Technology Corporation (MCC), a special type of university/industry R&D collaboration. Similarly, other collaborative R&D activities have been founded in the 1980s at Stanford University, MIT, Arizona State University, the University of North Carolina, and numerous other U.S. universities. The essential features of these collaborative university/industry R&D alliances are the following:

1. They are funded in large part by contributions from private firms;
2. They are located on or near a university campus or research park; and
3. University researchers (faculty and graduate students) collaborate closely with R&D workers in private firms in conducting research and in transferring the resulting technology to the private firms that are members of the R&D consortium.

Usually a collaborative R&D venture focuses on a specialized field, like microelectronics. An example would be the Engineering Excellence Program at Arizona State University. The key question is, How do collaborative university/industry R&D alliances encourage high technology development?

A related issue, which also concerns collaborative university/industry R&D alliances, is how they may create problems for the research university. For example, does collaborative R&D limit the free flow of scientific communication? Does it lead to an overemphasis on applied research at the expense of basic research? Does it lead the university to emphasize certain research topics, like microelectronics and biotechnology, at the expense of others, like the humanities? Here the question is, What are the advantages and disadvantages to the research university of participation in a collaborative university/industry R&D venture?

THE TREND TO COLLABORATION

In the United States in recent years, a trend has occurred from pure competition among high technology firms to certain types of collaborative relationships. The collaborative university/industry R&D venture is one illustration of this trend. Other outcroppings of collaborative relationships such as venture capital arrangements (which call for a partnership between the entrepreneurial firm and the venture capitalist), university research parks, and technology licensing agreements have been noted. Undoubtedly this trend toward collaborative relationships among U.S. firms has been inspired by concern over international competition. How has a possible trend toward collaborative arrangements, and the concern with international competition that inspires it, encouraged high technology development?

GOVERNMENT POLICIES

In addition to government policies that improve research universities and that encourage university/industry exchanges, what lessons can be learned about the effect of other public policies encouraging high technology development? Examples might be policies concerning taxes, labor unions, trade protection, and venture capital. How do national, state, and local government policies encourage high technology development? Here the wide range of international cases of technopoleis is useful in understanding the impacts of various public policies on high tech development.

ENTREPRENEURSHIP

What role does entrepreneurship play in high technology development? In Silicon Valley, certain early entrepreneurs served as important role models for later entrepreneurs in microelectronics. The mass media promote this entrepreneurial fever through the considerable attention they give to such individuals' success stories. Now that the bloom seems to be somewhat gone from Silicon Valley, is such entrepreneurship diminishing? How important is a past tradition of business risk in leading to entrepreneurship? In the case of technopoleis, we need to find out.

COMPETITION AND COOPERATION

A fascinating paradox has emerged—the paradox of competition and cooperation in the technopolis. On the one hand, intense competition takes place among and between universities, companies, cities, and states, and public and private-sector entities. On the other hand, cooperation is essential for a technopolis to develop. How do different technopoleis resolve this paradox?

Clearly, one important lesson has emerged: Those who manage technology creatively and innovatively will reap the benefits of sustained economic growth. The technopolis models presented in this volume describe how various areas are implementing this lesson. To facilitate an understanding of these models, the book is divided into three parts. Part I presents models of technopoleis developing outside the United States: Tsukuba Science City and Osaka in Japan; Beijing, China; Cambridge, England; and Sophia-Antipolis in France.

Part II examines technopoleis in the United States; mature technopoleis in Silicon Valley in California and around Route 128 in Massachusetts; developing technopoleis in the Austin/San Antonio Corridor in Texas and around Rensselaer Polytechnic Institute in Troy, New York; and an emerging technopolis in Phoenix, Arizona.

Part III explores issues and concerns that span technopoleis wherever they may develop. These include the role of basic science, the process of marketing the technopolis, entrepreneurship education, and the ramifications of intellectual property.

What these papers show is that developing and sustaining a technopolis is always a dynamic and creative, although sometimes uncertain and risky, process. They begin to address the questions and issues raised here, and in the process provide direction for further research and analysis. Our hope is that by presenting these papers we contribute to a better understanding and implementation of the technopolis concept.

Part I

INTERNATIONAL PERSPECTIVES

Chapter 1

BUILDING A JAPANESE TECHNOSTATE
MITI's Technopolis Program

Sheridan Tatsuno

In 1980 the Ministry of International Trade and Industry (MITI) announced the technopolis concept, an ambitious plan to build a Japanese technostate of research cities dispersed throughout the country. Based on Silicon Valley, California, and Tsukuba Science City, Japan, these technopoleis are designed to become centers of Japanese scientific and technological research in the twenty-first century. They will feature research universities, science centers (technocenters), industrial research parks, joint R&D consortiums, venture capital foundations, office complexes, international convention centers, and residential new towns.

Since passage of the technopolis law in 1983, prefectural governments have made significant progress in rallying local industries and universities around their technopolis plans. They have prepared twenty-year development plans, formed 255 R&D consortiums, and begun construction of highways, airports, industrial parks, and new towns. MITI has approved twenty regions and is currently reviewing six more sites. (See Figure 1–1.) In May 1986 the Japanese government approved MITI's Regional Research Core Concept, which calls for establishing research centers in twenty-eight regional cities. This research city policy is MITI's response to the rapid exodus of manufacturing plants overseas, better known as the "hollowing out" (*kudooka*) of the Japanese economy, which has been caused by the rapid yen appreciation.

This chapter examines the following developments in the technopolis program and Japanese high tech infrastructure policies:

Figure 1-1. Technopolis Areas in Japan.

1. Chronology of the technopolis program (1980 through 1986),
2. MITI's Regional Research Core Concept,
3. The status of four leading technopoleis (Okayama, Hiroshima, Ube, and Kumamoto),
4. The activities in the twenty technopolis regions (R&D facilities, information centers, joint R&D programs, land development projects, and large-scale infrastructure),
5. The new Y3.6 trillion ($22.7 billion) pump-priming package recently announced by the Japanese government.

Japanese regions, faced with declining exports due to the "yen shock," are seeking ways to reposition their industries to meet the growing Asian challenge. They are moving up the technology ladder to higher value-added products and services and promoting creative research. The technopolis program is the centerpiece of this new Japanese industrial strategy.

CHRONOLOGY OF THE TECHNOPOLIS PROGRAM

Since 1980 MITI has worked closely with prefectural governments to develop the technopolis program. In 1981 thirty-eight of the forty-seven prefectures volunteered to become technopolis sites, forcing MITI to establish a screening process. In March 1983 after conducting detailed surveys overseas and discussions, MITI issued the following selection criteria:

1. Completion of technopolis construction by 1990,
2. Development areas of less than 500 square miles,
3. Location of the technopolis within a thirty-minute commute of a regional city (parent city) of 150,000 people.

These guidelines were incorporated into the technopolis law, which was passed by the Japanese diet in April 1983. Under this law, the technopolis regions are promoting five major development policies:

1. Integrated development of industry (*san*), research universities (*gaku*), and housing (*juu*),
2. Close ties between the technopolis and its parent city (*botoshi*),
3. Balanced development between new high tech industries and technological upgrading of existing industries,
4. Transfer of high technology to existing industries (transfer R&D) and creative research in frontier fields (frontier R&D),
5. Regional uniqueness in high tech research and industrial development.

In 1983, nineteen regions were given preliminary approval by MITI, which requested detailed development plans, as shown in Table 1–1. Since 1984 MITI has given final approval to twenty regions (Nagasaki was added to the list) and is considering six additional sites. The approved technopolis regions are eligible for national tax incentives and special depreciation allowances.

MITI'S REGIONAL RESEARCH CORE CONCEPT

Until early 1986 the technopolis program was primarily aimed at relocating high tech manufacturing from Tokyo and Osaka to the regions. Due to the yen

Table 1-1.   Chronology of the Technopolis Program.

Date	Activity
03/80	Technopolis concept announced in MITI's visions for the 1980s
07/80	Survey by local industrial city concept study group
06/81	Survey by MITI's Technopolis '90 Committee
03/82	Technopolis basic plans established
06/82	Interim report by Technopolis '90 Committee
05/83	Technopolis development plans completed by prefectures
07/83	Technopolis law enacted by Japanese diet
10/83	Development guidelines announced by MITI
03/84	First-round approval of development plans for Niigata, Toyama Shizuoka, Hiroshima, Yamaguchi, Kumamoto, Oita, Miyasaki, and Kagoshima prefectures
05/84	Akita and Tochigi prefectures approved
07/84	Hokkaido prefecture approved
08/84	Okayama prefecture approved
09/84	Fukuoka and Saga prefectures approved
03/85	Nagaskai prefecture approved
04/85	First Technopolis Symposium held in Tokyo ("Technopolis: MITI's 21st Century Super-Vision")
08/85	Aomori prefecture approved
09/85	Hyoog prefecture approved
04/86	Cabinet ministerial meeting on "yen shock" comprehensive countermeasures (promotion of technopolis concept)
05/86	Miyagi and Fukushima prefectures approved
05/86	Private investment law (regional R&D core concept) passed by Japanese Diet
09/86	Economic stimulation package adopted by Japanese cabinet (technopolis concept as vehicle for rural and regional development promotion policy)
11/86	Second Technopolis Symposium held in Kumamoto City
12/86	Miyagi (Sendai) and Fukushima (Koriyama) prefectures approved

Source: MITI, 1986.

shock, however, Japanese companies have been shifting their production facilities overseas instead of to the technopoleis. To counter this offshore movement, MITI developed the Regional Research Core Concept to promote the regionalization of high tech research and to strengthen local R&D facilities. This program, which was passed by the Japanese Diet in May 1986, promotes four types of research facilities:

1.   Open experimental research institutes for joint industry/academic/government research,

Figure 1–2. Research Core Project Program.

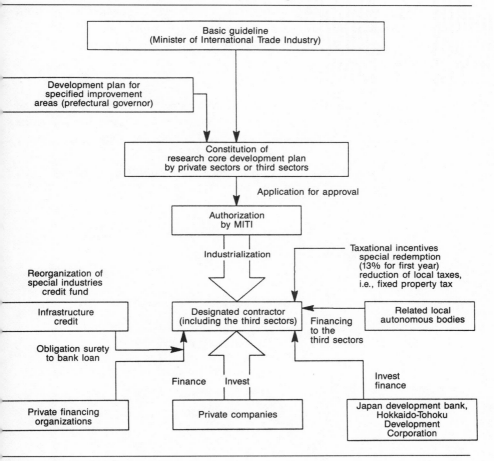

Source: MITI, 1986.

2. New research training and education facilities,
3. Conference halls, exhibition halls, and data base systems for improved access to technical information,
4. Venture business incubators.

Using Silicon Valley as a model, MITI is also promoting support facilities, such as business complexes for service industries, venture capitalists, legal and accounting professionals, and think tanks. The private sector will be encouraged to participate in developing local research core plans according to MITI guidelines, as shown in Figure 1–2. Research core cities will be eligible for tax benefits, insurance guarantees, and financing loans from the Japan Development Bank (JDB) or Hokkaido-Tohoku Development Corporation.

MITI is studying twenty-eight regions for the Regional Research Core Program, as shown in Figure 1–3. As listed in Table 1–2 eighteen of the research core projects are also technopoleis. The following is a short description of three research core projects:

1. Twenty-first Century Plaza (Miyagi Prefecture, #5, Figure 1–3). This Y34 billion ($212 million) complex will feature two zones: a Technoculture Zone consisting of research institutes, information centers, and venture

Figure 1–3. Main Candidates for Research Core Projects.

1-Eniwa (Hokkaido)
2-Aomori
4-Akita
3-Morioka
6-Yamagata
5-Izumi (Miyagi)
14-Nagaoka (Niigata)
15-Toyama
7-Koriyama (Fukishima)
13-Nagano
8-Utsunomiya (Tochigi)
17-Okayama
9-Katsuta (Ibaraki)
18-Hiroshima
19-Ube (Yamaguchi)
10-Kisarazu (Chiba)
11-Kawasaki (Kanagawa)
22-Kurume (Fukuoka)
23-Saga
12-Kofu (Yamanashi)
24-Nagasaki
16-Kishiwada (Osaka)
25-Kumamoto
26-Oita
20-Matsuyama (Ehime)
21-Tokushima
27-Miyazaki
28-Kagoshima

Source: MITI, 1986.

Table 1-2. Candidate Sites for Research Core Projects.

Name of Project	*Location*
1. Eniwa High Complex City	Eniwa City, Hokikaido
2. Aomori Future Park	Aomori City, Aomori Prefecture
3. Technopolis Support Core	Morioka City, Iwate Prefecture
4. Akita Core City Key Area Industrial Support Development Project	Akita City, Akita Prefecture
5. Twenty-first Century Plaza ($213 million)	Izumi City, Miyagi Prefecture
6. Yamagata International Industrial Communication Plaza	Yamagata City, Yamagata Prefecture
7. Koriyama Regional Technopolis Industrial Support Functions Development Project	Koriyama City, Fukushima Prefecture
8. Technopolis Center	Utsunomiya City, Tochigi Prefecture
9. Hitachi-Naka International Bay Park City "Business Pleasure Hitachi-Naka"	Katsuta City, Ibaragi Prefecture
10. Kazusa New Development and Research City Plan "Academia Park"	Kisarazu City, Chiba Prefecture
11. Kanagawa Science Park	Kawasaki City, Kanagawa Prefecture
12. Science Park (Twenty-first Century Industrial Park	Kofu City, Yamanashi Prefecture
13. Technoculture Zone	Nagano City, Nagano Prefecture
14. Technovalley Intelligent Core Project Plan ($100 million)	Nagaoka City, Niigata Prefecture
15. Toyama Intelligence Corridor	Toyama City, Toyama Prefecture
16. Cosmopolis Plan (Research Park)	Kishiwada, Izumi-Sano, Izumi Cities, Osaka Prefecture
17. Okayama Triangle D&R Project Plan	Okayama City, Okayama Prefecture
18. Hiroshima Central Technopolis Innovation Park	Higashi-Hiroshima City, Hiroshima Prefecture
19. Ube New City Techno Center	Ube City, Yamaguchi Prefecture
20. Tokushima Prefectural Industrial Research Core	Tokushima City, Tokushima Prefecture
21. Techno Plaza Ehime	Matsuyama City, Ehime Prefecture
22. Kurume Techno Research ($125 million)	Kurume City, Fukuoka Prefecture
23. Saga Research Core	Saga City, Saga Prefecture
24. Creative Area Project	Nagasaki City, Nagasaki Prefecture
25. Kumamoto Creative Area	Kumamoto Prefecture
26. Oita Intelligent Zone	Oita City, Oita Prefecture
27. Miyazaki Sun-Tech Park	Miyazaki City, Miyazaki Prefecture
28. New City Center	Hayato Cho, Kagoshima Prefecture

Source: MITI, 1986.

business incubators, computer shops, and an International Convention Center Zone. Construction will begin in fiscal 1987.

2. Technovalley Intelligent Core (Niigata Prefecture, #14, Figure 1–3). Located in the Nagaoka Techno-Valley Technopolis, this Y16 billion ($100 million) project will consist of the Nagaoka Regional Technology Development Promotion Center, International Communication Plaza, Education Plaza, and venture business incubators. Construction will begin in 1988.

3. Kurume Techno Research Park (Fukuoka Prefecture #22, Figure 1–3). This Y20 billion ($125 million) technopolis project will feature a research center, industrial parks, and venture business incubators. Construction will begin in fiscal 1987.

STATUS OF FOUR LEADING TECHNOPOLEIS

Okayama

Okayama's Kibi Highland Technopolis (#17, Figure 1–3) has the goal of becoming the biotechnology center of Japan. In 1985 Okayama University formed the Biotechnology Research Lab, which has twenty researchers pursuing fermentation research. In 1986, 300 industry and government researchers established the Biotechnology Research Association with Okayama University's Pharmaceutical Department to build a Biotechnology R&D Center in the Kibi Highland Technopolis by 1990. Currently, Hayashibara Biochemical (which supplies 50 percent of the world's supply of interferon) is building an interferon production plant and developing semiconductor bioresists with Matsushita Electronics. The prefecture recently built a $10 million Life Science Center next to the Japan Industrial Worker Rehabilitation Center. The Matsushita video plant nearby is specially designed for physically handicapped workers. Located an hour west of Osaka on the bullet train, Okayama will be the crossroads to Shikoku Island when the Seto Island sea bridge is completed in 1988. The new Okayama Airport is being built on a plateau within the Kibi Highland Technopolis.

Hiroshima

The Hiroshima Central Technopolis (#18, Figure 1–3) is focusing on five new strategic industries: electronics, mechatronics, new materials, biotechnology, and new energy development. In 1982 the engineering and science departments of Hiroshima University were moved to the new Kamo Science City, which is located in the hills twenty-five miles east of Hiroshima City. A new biotechnology Center is currently under construction. The prefecture will complete its master plan this year and begin constructing a Technoplaza (a venture business incubator offering training programs and technical advice) in 1989. Expressways and roads are being built to improve access, and a new bullet train terminal will be completed in 1988. The new Hiroshima Airport will be built by 1993. The prefecture is also preparing a Marinopolis plan to promote marine research.

Yamaguchi

The Ube Phoenix Technopolis (#19, Figure 1–3) is promoting a broad range of fields, including electronics, mechatronics, polymers, biotechnology, fine chemicals, marine resources, energy development, and software. In April 1987 Yamaguchi University opened a $2.6 million research facility and international exchange center. Recently, the university established a Mechano Technology Center for graduate students and professors, who will be free to conduct their own creative research outside the constraints of the university research system. Yamaguchi prefecture, which plans to build a new research city of 14,000 by 1991, was given a boost when MITI selected Ube as a Regional R&D Core City this year. The new city will cost Y160 billion ($1 billion), of which half will be spent on sewers and utilities. Currently, MITI, the Ministry of Transportation (MOT), Ministry of Agriculture, Forestry, and Fisheries (MAFF), and local companies are planning marine R&D laboratories for Ube's Marinovation Program.

Kumamoto

Kumamoto prefecture's Governor Morihiko Hosokawa is one of the strongest advocates of the technopolis program. Patterned after the Research Triangle in North Carolina, the Kumamoto Technopolis (#25, Figure 1–3) will focus on automation, biotechnology, computers, and data processing. The region is located at the heart of Kyushu, better known as Silicon Island because it produces 40 percent of all Japanese semiconductors. NEC, Matsushita, and other IC makers are located in the area. In April 1985 the Applied Electronics Research Center was opened next to the new Kumamoto International Airport to conduct semiconductor-related research. A new IC start-up, Digital

Design Systems (DDS), is developing gate arrays and PLDs at the center. In 1986 the technopolis center was completed nearby to conduct training classes and provide on-line technical information to local companies. Kumamoto, which introduced a private videotex KINGS system in 1985, was recently designated a Teletopia site by the Ministry of Posts and Telecommunications (MPT). The region also plans to develop a Software Forest and a Biotechnology Forest to conduct basic and applied research.

TECHNOPOLIS R&D CENTERS

Since 1982 Japanese technopoleis have strengthened their public and private research institutes to raise the technological level of local industries (called *level-up*), transfer technology from Tokyo and Osaka, and conduct independent research. Twenty-one research facilities have already been built, and there are plans for nineteen others beginning in fiscal 1986. Table 1–3 lists sixteen of the twenty-one new facilities.

Table 1-3. New Technopolis R&D Centers.

Region	Research Facility	Date
Hakodate	Hokkaido Industrial Technology Center	08/86
Akita	Akita Technical Center	10/82
Nagaoka	Nagaoka Technology Development Center	08/84
Hamamatsu	Mechatronics Research Institute	03/84
Toyama	Toyama Technology Center	07/86
Okayama	Kibi Highlands New Science House	10/85
Okayama	Biotechnology Research Laboratory	
Hiroshima	Innovative Technology Center	04/85
Hiroshima	Technoplaza	FY89
Ube	Yamaguchi Mechatronics Technical Center	03/87
Ube	Ube Techno-Center (MITI R&D Core Program)	FY87*
Kurume-Tosu	Kurume Techno-Research Park (MITI R&D Core Program)	FY87*
Kenhoku-Kunisaki	Oita Prefecture Advanced Technology Development Research Institute	06/84
Kumamoto	Applied Electronics Research Center	03/85
Miyazaki	Miyazaki Joint R&D Center	10/84
Kokubu-Hayato	Kagoshima Fine Ceramics Products R&D Institute	10/84

Source: MITI, 1986.
*Begin construction.

ON-LINE INFORMATION CENTERS

Prefectural governments are building technocenters, which will provide technical information through on-line data base networks, conduct training programs on data processing and communications systems, and act as information clearinghouses. To date, twenty-three centers have been created; twelve others will be established in fiscal 1986. Table 1–4 lists only the major centers. In addition, Japan Techomart, Inc., a newly formed technical information service, has opened local offices in Hamamatsu, Toyama, Hiroshima, Kurume-Tosu, Kumamoto, Sendai, Yamagat, Asama, and Ehime to link these cities with Tokyo and Osaka.

TECHNOPOLIS FOUNDATIONS

In each region, technopolis foundations are being established to promote technopolis construction and introduce advanced technologies to local companies. These foundations act as research debt guarantors and organizers of training seminars and trade shows. To date, they have achieved the following:

1. Underwritten Y1.3 billion ($7.9 million) in low-interest loans and interest subsidies to seventy-three companies nationwide for new technology and product development (as of fiscal 1985),
2. Sponsored 363 engineering and executive training programs in eighteen regions (as of fiscal 1986),

Table 1–4. Technopolis On-Line Information Centers.

Region	Center	Opened
Hamamatsu	Information Technology College	04/85
Hamamatsu	Local Technology Development Center	03/85
Toyama	Toyama Information and Training Center	10/84
Toyama	Toyama Technical Exchange Center	05/85
Hiroshima	Kure Personal Computer Center	04/85
Ube	Mine Industrial Technology Center	03/85
Nagasaki	Nagasaki Prefecture Venture Business Information Center	04/85
Oita	Oita Soft Park	03/85
Kagoshima	Data Processing Training Center	04/86
Kumamoto	Kumamoto Technopolis Center	09/86

Source: MITI, 1986.

3. Conducted studies of new social systems for the technopolis regions,
4. Introduced technical information services, such as JOIS and PATOLIS, to local companies,
5. Formed researcher talent data bank services,
6. Sponsored numerous technopolis fairs and high technology trade shows.

Generally, the boards of these foundations consist of the governor, local mayors, industry leaders, university professors, media executives, labor union leaders, and others. These high tech chambers of commerce tend to be more internationalist than their local constituencies, and they welcome foreign investments in their regions.

REGIONAL JOINT R&D PROJECTS

The technopolis regions are actively promoting joint research projects and interindustry exchanges to raise the technological standards of local companies. By fiscal 1986 there were 255 joint research projects among the eighteen regions, involving local companies, universities, public research institutes, and national testing laboratories. These projects focus on applied research and technology transfers from Japan's National R&D Laboratories. However, the Regional Frontier Technical Development Project, financed by MITI's Small and Medium Enterprises Agency, focuses on region-specific research. Table 1–5 lists the major R&D projects underway.

The following projects will have a major impact on the semiconductor industry:

1. Utsunomiya—automotive electronics and production technology,
2. Kagoshima—fine ceramics production research with Kyocera and over 100 regional ceramics makers,
3. Hamamatsu—optoelectronics for factory and office automation,
4. Nagasaki—electronic controls for new material production,
5. Oita—vision and other sensor technology,
6. Kumamoto—semiconductor applications.

LAND DEVELOPMENT PROJECTS

In order to attract high tech industries, the technopoleis are actively building new industrial parks located near airports, expressways, and bullet train stations.

Table 1–5.   Joint Government/Industry/University Research Projects.

Region	Project	Research Focus
Hakodate	Hakodate Regional Technology Promotion Committee	Marine resources and equipment Food processing
Aomori	Absorptive Ceramics Development Project	Advanced applications
Akita	Akita Prefecture Technology Frontier Project	Automatic selection system for multicolor picture processing
Nagaoka[a]	Regional Technology Development Committee	Precision machinery and information control
Utsunomiya[a]	Applied Electronics Research Association (six IC companies), Fine Ceramics Research Association (eleven materials companies), CAD/CAM Research Association (eight carmakers)	Production process technology for next-generation automotive design, bodies, and electronics (see *JSIS Bulletin*, March 5, 1986)
Hamamatsu[a]	Applied Phototechnology Produciton Process R&D Solar Research Laboratory Medical Equipment Technology Research Laboratory	Optoelectronics for factory automation Solar batteries and cells Artificial organs and medical
Toyama[a]	Aluminum Research Association Hokuriku Machining Center Research Association	Flexible manufacturing system (FMS) response technology; plastics molding simulation system
Nishi-Harima (Kobe area)	Hyogo Mini-Frontier Project	New ceramics and amorphous metal conferences
Kibi	Biotechnology R&D Association	Fermented food technology
Highlands (Okayama)[a]	Precision Process Association	Home care robots and precision appliances for handicapped and elderly
Hiroshima[a]	Advanced Material Processing Technology Association	Carbon fibers
Ube (Yamaguchi)	Ube New frontier Project Fine Stone Plate Research Association; Fly Ash Research Association	Automated die manufacturing technology; fine stone plate; fly ash

Table 1–5. Continued

Region	Project	Research Focus
Kagawa	New Ceramics Products Development Research Association	Fine ceramics
	Applied Microbiology Technology Research Association	Biotechnology
Kurume-Tosu[a]	New Ceramics Material and Products Research Association	Fine ceramics
Kan-Omurawan (Nagasaki)[a]	New Materials Research Association; Nagasaki Advanced Technology Development Council	Applied electronics control technology for new materials
Kenhoku-Kunisaki (Oita)[a]	Oita Prefecture Technical Exchange Plaza; Oita Personal Computer Research Association	Sensor applications, especially vision sensors
Kumamoto[a]	Applied Semiconductor Technology Research Association	Semiconductor applications
	Biotechnology Research Research Association	Biomedicines
Kokubu-Hayato[a] (Kagoshima)	New Ceramic Products Research Association; Kagoshima Natural Resources Development Council	Fine ceramics applications Natural resources

Source: Dataquest, Inc., 1987.

[a]Local Frontier Technology Development Project financed by MITI's Small and Medium Enterprises Agency.

By fiscal 1985 about 4,360 hectares (10,770 acres) of land had been prepared for fifty-eight industrial parks; another 2,470 hectares (6,095 acres) will be prepared for sixty-one more industrial parks from fiscal 1986 and beyond.

LARGE-SCALE INFRASTRUCTURE PROJECTS

Large-scale infrastructure projects are also being built to support the new technopolis research centers and industrial parks. The technopoleis are busy constructing new technoroads, water and sewer systems, highways, bullet train terminals, airports, recreational parks, housing, telecommunications networks, community centers, and other public facilities. MITI's goal is to have the prefectures build and finance their basic infrastructure by 1990. Four areas of activity are particularly noteworthy:

1. Regional electric power companies (Kyushu Electric Power and Tohoku Electric Power) that have benefited from the windfall of cheap oil are expanding their electricity networks and aggressively promoting technopolis construction.
2. MITI, the Ministry of Construction (MOC), and the Ministry of Transportation (MOT) are implementing a Commuter Airport Program to link the technopoleis and regional cities to Tokyo and Osaka.
3. MITI and the Ministry of Posts and Telecommunications (MPT) are competing to introduce new telecommunication networks (MITI's New Media Community concept and MPT's Teletopia Concept).
4. Japan National Railway (JNR) is planning to extend the bullet train (Shinkansen) to southern Kyushu and Osaka to Nagaoka.

JAPANESE GOVERNMENT UNVEILS A PUMP-PRIMING PACKAGE

Recently, the Japanese government announced a set of fiscal and monetary policies designed to counteract the recessionary effects of the yen shock and to boost domestic demand as recommended in the Maekawa Report. These measures will strongly affect industrial development in the technopoleis. The specific policies include

1. Lowering the prime rate to 3 percent,
2. Offering special low-interest loans and easing industrial regulations for ailing manufacturers, and
3. Increasing public works spending and encouraging more private investment.

In September 1986 the Japanese government unveiled a comprehensive investment program that calls for Y3.636 trillion ($22.7 billion) in increased public and private spending. The Japanese government has allocated Y3 trillion ($18.75 billion) for housing loans, highways, sewers, and other social infrastructure. About Y133 billion ($830 million) in new construction bonds will be issued to finance this increased public spending. Under the new private-sector vitality law (Minkatsu law), private industry will be encouraged to invest Y636 billion ($3.98 billion) in urban redevelopment, international trade fairs, and other commercial projects. The Economic Planning Agency expects this public/private investment program to have a net stimulative effect of Y4.9 trillion ($30.6 billion), or 1.5 percent of the fiscal 1986 GNP.

Local prefectures will spend about Y800 billion ($5 billion) of this total package, or about $106 million per region. Because technopolis sites are located in twenty-six of the forty-seven prefectures, the technopoleis will benefit from $2.8 billion in public and private investments in 1986.

Although the United States is urging Japanese consumers to spend more on foreign imports, the new economic package emphasizes productivity-boosting investments in public works and high tech infrastructure. Due to the emphasis on private investment, there are many investment opportunities for multinational companies. In particular, the Japanese government is promoting the following types of public and private investments.

LARGE-SCALE PROJECTS

Japanese ministries and local governments are promoting large-scale projects to boost local demand and create new jobs. Efforts include

1. Large-scale public facilities, such as the Kansai International Airport (Y1 trillion/$6.25 billion), Trans-Tokyo Bay Highway (Y1.15 trillion/$7.19 billion), and Makuhari New Metropolis (Y1 trillion/$6.25 billion);
2. Over Y110 billion ($688 million) in investments by electrical power and gas companies in fiscal 1986 and front-loading of Y200 billion ($1.25 billion) of their orders in fiscal 1987; and
3. An extra Y50 billion ($313 million) in capital investments by telecommunications companies in the last half of fiscal 1986.

PRIVATE-SECTOR INVESTMENTS

Under the private investment law, the private sector is being encouraged to reinvest in the domestic economy, not only overseas. Efforts include

1. Relaxing of Japanese Housing Corporation loan provisions to allow for larger homes and more rental housing;

2. Promotion of plant construction and expansion in technopoleis and other designated industrial parks;
3. Construction of international trade fair and conference halls, disaster prevention systems, on-line data base systems, intelligent buildings designed with telecommunication networks, underground shopping malls, land rezoning, and regional development;
4. Special urban and port development zones, such as the Kansai Culture and Science Research City in the Osaka area or the Port Future 2000 (Minato Mirai 2000) in Yokohama, which will be eligible for financial assistance, low interest loans, and special tax depreciation (20 percent).

NEW RESEARCH CENTERS

Besides technopoleis, other high technology infrastructure projects are being planned in major cities. These include

1. Regional R&D core cities consisting of new research centers, joint R&D projects, and venture business incubators located in twenty-eight regions;
2. Development of new high tech research cities in the Tokyo and Kansai regions, as shown in Figures 1–4 and 1–5; and
3. Construction of regional software development centers by major Japanese electronics companies.

DEREGULATION

As in the United States, privatization and deregulation are seen as ways to increase economic efficiency and reduce budget deficits. This is being done by

1. Deregulation of the aviation industry, gas station construction and services, and charter bus and taxi licensing; and
2. Privatization of Japan National Railway (JNR) and development of underutilized land for office and commercial use.

ASSISTANCE TO SMALL BUSINESSES

The Japanese government is assisting small businesses that have taken the brunt of the yen shock. Efforts include

Figure 1–4. Tokyo High Tech Cities.

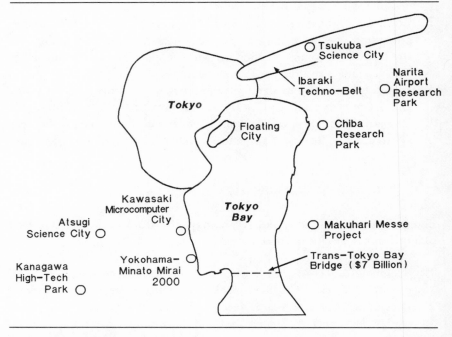

Source: Dataquest, Inc., January 1987.

1. Temporary financial aid, credit insurance, and loan programs for company restructuring, research, and repositioning;
2. Government bank lending (Japan Development Bank, Long-Term Credit Bank of Japan, and so forth);
3. Jointly sponsored excess equipment sales programs; and
4. Subcontractor protection policies to prevent contracting companies from taking advantage of subcontractors.

These measures are being coordinated by the national and local governments in conjunction with the private sector.

CONCLUSION

Since fall 1986 MITI has introduced major policy changes in the technopolis program, which are rapidly being implemented by local governments. Initially,

Figure 1–5.   Kansai High Tech Cities.

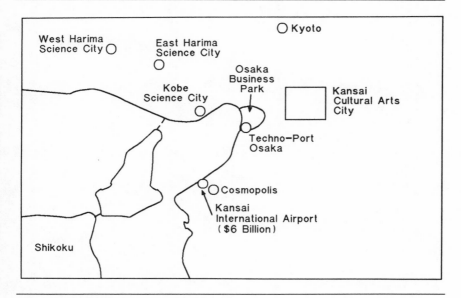

Source: Dataquest, Inc., 1987.

many technopolis plans looked like idealistic paper dreams, but the prefectures seriously view them as long-term regional business plans. There is no guarantee that all of the technopoleis will succeed, but few options are available given the yen shock. Industries must move up the technology ladder as quickly as possible if they are to remain internationally competitive with the emerging Asian countries in the 1990s.

The regionalization of high tech research in Japan is a major policy shift that will strongly affect the global economy. Creative research and innovative new products will emerge from these technopoleis and research core cities by the mid-1990s, if not sooner.

Chapter 2

TECHNOPOLIS OSAKA
Integrating Urban Functions
and Science

Keisuke Morita and Hiroshi Hiraoka

The Osaka Metropolitan Area (OMA), facing Osaka Bay, boasts more than 1,500 years of history. With the synergistic development of closely located core cities, including the economic center of Osaka, the cultural centers of Kyoto and Nara and the trade center of Kobe, this area has evolved into Japan's second-largest hub of economic and cultural activities after the Tokyo Metropolitan Area. Through the analysis of high tech businesses, research institutes, and universities established in the OMA, this chapter elucidates the requirements for a technopolis conducive to regional development and introduces a new approach to Technopolis Osaka, which seeks to fully integrate existing urban functions.

The Kinki region, located roughly at the center of Japan, comprises an area of 37,247 km^2 (14,381 sq mi). Mountains account for 70 percent of the area. The densely inhabited district comprises approximately only 5 percent of the region, or 1,918 km^2 (741 sq mi), but contains 70 percent of the regional population.

Historically, Japan's first unified state was constructed in the southern part of Nara around the sixth century. The political center moved to Osaka, again to Nara and then to Kyoto. These cities grew as cultural, commercial, and art centers. Kyoto in particular has been the center of culture and art since the eighth century. During the seventeenth and through nineteenth centuries, Osaka prospered as Japan's economic hub, garnering 70 percent of the goods

23

and boasting a maximum population of 400,000. Since the late nineteenth century Kobe has developed primarily through international trade.

With the construction and improvement of railroads and highways, the Kinki region, consisting of Osaka, Kyoto, Fukui, Shiga, Mie, Nara, Wakayama, and Hyogo prefectures, with Osaka, Kyoto, and Kobe serving as its core cities, has evolved into a large metropolitan area with a total population of 22 million. The agglomeration of urban functions in the region, surpassed only by that in the Tokyo metropolitan area, involves some 1,280,000 business establishments, 9.9 million employees, Y48.9 trillion ($362 billion) in annual shipments and Y211.4 trillion ($1.56 trillion) in annual wholesale/retail sales. Expressways and the Shinkansen trunk line, which constitute national land traffic axes, pass through Kyoto, Osaka, and Kobe, linking them to eastern and western Japan.

The OMA consists of sixty-three cities and sixty-three towns/villages located within roughly 50 km (31 mi) of Osaka City. The area comprises 4,201 km^2 (1622 sq mi) where some 16.6 million people reside. Its nucleus, Osaka City, embraces an area of 212 km^2 (81.5 sq mi) and has a residential population of roughly 2,640,000. Approximately 1,250,000 people daily commute into Osaka City for work, school, and other purposes, while 240,000 people flow out. The daytime population increases by 1,010,000 to some 3,650,000. The OMA is home to roughly 915,000 business establishments, which absorb some 7.4 million employees, 65 percent of the total number of employees in the Kinki region. In Osaka City, business establishments number 274,000; employees total 2,470,000. Particularly the four central wards, with a combined area of approximately 20 km^2 (7.72 sq mi, 9 percent of the entire city area), comprise a highly dense central business district (CBD) where over 1 million people work during the daytime.

The Kinki region faces two major tasks. One is to develop and reinforce its administrative function, which in recent years has tended to be concentrated in the Tokyo metropolitan area. The Kinki region must develop and reinforce this function to achieve better national security and well-balanced development nationwide. The other task is to upgrade traffic and communications networks to ensure a synergistic effect of administrative and various other urban functions.

In line with these regional objectives, Osaka City confronts the following tasks:

1. To develop and reinforce administrative functions for more comprehensive and integrated economic development in the region;

2. To upgrade the information and communications infrastructure to accommodate the information society; and
3. To improve overall urban amenities and attract diverse people.

LEADING PROJECTS IN OMA

As information, communications, and traffic technologies advance, the nations of the world are becoming increasingly interdependent in the economic, social, and cultural realms. Japan, aiming at becoming a technology- and knowledge-oriented country, will fulfill an important role in the international society of the twenty-first century. Particularly crucial in this task is the role of the OMA, which boasts rich concentrations of economic and social functions and traditional culture.

Essential to developing a regional economy is the improvement of the urban infrastructure, on which will rest the economic and cultural activities of the OMA and Japan in the coming century. In the OMA, two major projects currently underway, are expected to greatly stimulate future development (see Figure 2–1): Kansai International Airport and Kansai Science City. Kansai International Airport, Japan's first twenty-four-hour-access offshore airport to accommodate international economic activities, will be the air traffic gateway not only to the OMA but to all of Japan. Kansai Science City is intended to serve as a source of creative knowledge and information incorporating Japan's cultural tradition, so that Japan can create technology and information.

At the existing Osaka International Airport, located in the densely populated nothwestern area of Osaka City (Figure 2–1), noise problems have led to the banning of all takeoffs and landings during ten hours at night and to the restriction of the number of jet takeoffs and landings to 200 daily. As a result, airlines of only eight countries are now allowed to use this airport, with more than thirty other nations on the waiting list. Osaka International Airport is thus one of the most restricted major airports in the world.

In view of the importance of air traffic to international economic activity, a twenty-four-hour-access international airport is an essential to Japan's future. The Kansai International Airport, which will offer the first around-the-clock service in Japan, will be critical in serving not only the OMA but all of Japan with an international air traffic network. The airport will be constructed on an artificial island in Osaka Bay (5 km or 3.1 mi offshore). Following land reclamation to create an island of 511 ha (1,262 acres), airport terminal facilities and a runway (3,500 m or 11,483 ft) will be constructed. A road-rail combined

Figure 2-1. The Osaka Metropolitan Area.

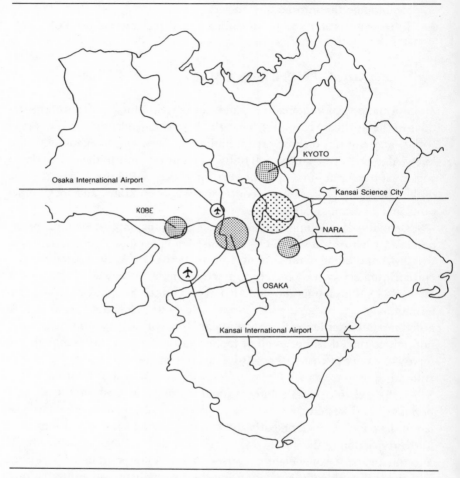

Source: Comprehensive Planning Bureau, Osaka Municipal Government, 1987.

bridge will link the island with the mainland. After opening, targeted for 1992, the airport is planned to serve 50,000 passengers and handle 1,800 tons of cargo with 160 takeoffs and landings daily. Future expansion to a total area of 1,200 ha (2,964 acres), with three runways, is also planned.

Kansai International Airport will enhance the high-speed mobility of business travel, both within Japan and abroad, and will bring tourists to the Kinki region to visit places of natural beauty and historic interest and to enjoy a dynamic urban culture. The following benefits are also expected:

1. Increased number of direct flights to more points in the world and concomitantly enhanced significance as a key airport in the international air traffic network;
2. Better conditions for providing the OMA with greater administrative capacity in the industrial, public, and academic sectors;
3. Increased opportunities of information exchange through international personal contact; and
4. Stimulate economic activity through increased volume and speed of air cargo handling.

The construction of the Kansai International Airport will cost Y1 trillion ($7.4 billion); the costs of all public works related to the airport (roads, railroads, and so forth) including construction of the airport itself, will amount roughly to Y4.3 trillion ($31.9 billion). It is estimated that this airport will serve 1.7 times as many passengers and handle about five times as much cargo as does the existing Osaka International Airport and will have an economic effect equivalent to Y600 billion ($4.4 billion) annually.

KANSAI SCIENCE CITY

Tsukuba Science City now serves as the hub of research in eastern Japan. As its western Japan equivalent, Kansai Science City is under construction on the Keihanna Hills, 30 km (18.6 mi) east of Osaka City. Kansai Science City will combine cultural, scientific, and ecological elements. Planned as clusters of research institutes, universities, housing, and other facilities, it will make full use of the natural topography of the Keihanna Hills, which remain rural.

A plan for the construction of Kansai Science City was proposed in September 1978. The prime minister decided to prepare a master plan for the city in November 1978 and the National Land Agency commenced investigation for that purpose in April 1979. In 1980 the Complex Plan and the construction of its center in the Keihanna Hills were decided on. In 1981 Kyoto, Osaka, and Nara prefectures worked out a pilot plan involving 2,500 ha (6.175 acres) in the Keihanna Hills. From April 1982 to March 1984 the National Land Agency, the Ministry of Agriculture, Forestry, and Fisheries, the Forestry Agency, the Ministry of International Trade and Industry, the Ministry of Transport, and the Ministry of Construction conducted investigations for comprehensive planning with regard to land utilization, traffic networks, housing, and the general living environment. In June 1986 the Kansai Research Institute was established for full-scale construction of the city, which will comprise eight clusters as described in Table 2–1.

Table 2-1. Kansai Science City.

Cluster No.	Cluster Name	Area (area for research facilities)	Functional Target	Facilities
1	Tanabe (Tanabe-cho, Kyoto pref.)	100 ha 247 acres (95 ha) 235 acres	Education, research	1. Doshisha University (1986) 2. Doshisha Women's College (1986) 3. Doshisha International Highschool (1980)
2	Minamitanabe-Komada (Tanabe-cho and Seika-cho, Kyoto pref.)	340 ha 840 acres (20 ha) 49 acres	Recreation residence	
3	Seika-Nishikizu (Seika-cho and Kizu-cho, Kyoto pref.)	490 ha 1210 acres	Research, information exchange (central cluster)	1. International Institute for Advanced Studies (established Aug. 1984) 2. Advanced Telecommunications Research Institute International (established March 1986) 3. Related facilities include: ATR Communication Systems Research Laboratories (basic research on intellectual communications system) ATR Interpreting Telephony Research Laboratories (basic research for automatic translation telephone) ATR Auditory and Visual Perception Research Laboratories (human-scientific basic research on visual-aural mechanism)

	Location	Area	Purpose	Institution
4	Kizu (Kizu-cho, Kyoto pref.)	740 ha 1828 acres	Research, leading-edge business	ATR Optical and Radio Communications Research Laboratories (basic research on optical and radio communications system)
5	Himuro-Tsuda (Hirakata City, Osaka pref.)	200 ha 494 acres (9 ha) 22.23 acres	Education, research, residence	Kansai University of Foreign Studies (1984)
6	Tawara (Shijonawate City, Osaka pref.)	130 ha 321 acres	Residence, recreation	
7	Kiyotaki-Muroike (Shijonawate City, Osaka pref.)	340 ha 840 acres	Recreation, education, research	Osaka Electro-Communication College
8	Heijo Palace Site (Nara City, pref.)	200 ha 494 acres	Research on cultural properties	

Source: Comprehensive Planning Bureau, Osaka Municipal Government, 1987.

THE OSAKA METROPOLITAN AREA AND OSAKA CITY

Of 218 foreign-affiliated firms located in the OMA in December 1985, 52 percent are in manufacturing (114 firms); 11 percent (twenty-four firms) in each of chemical and general machinery industries; and 7 percent (fifteen firms) in each of pharmaceutical and electrical equipment industries. Commercial and service fields constitute 42 percent (ninety-one firms) and 6 percent (twelve firms), respectively. Ninety firms (41 percent) are from the United States, followed by West Germany (twenty-six firms, 12 percent), Switzerland (eighteen firms, 8 percent), United Kingdom (sixteen firms, 7 percent), France (fifteen firms, 7 percent), and Hong Kong (nine firms, 4 percent). Most firms belong to western countries.

Foreign banks have been rapidly increasing in recent years. There is now a total of forty (including resident offices) foreign banks in Osaka: United States (7), Korea (7), France (6), Singapore (4), in addition to banks from some ten other countries.

As internationalization rapidly progresses throughout the entire OMA, international functions are also being reorganized, as evidenced by location changes of general consulates of foreign countries, from five in Osaka City and twelve in other districts in the 1970s to eleven and eight, respectively, in 1987.

The role of Osaka City in the OMA is changing. The area is undergoing postindustrialization, as evidenced by

1. Shifts from manufacturing to wholesale, retail, and service industries;
2. Shifts from basic material supply in the manufacturing industry to processing and assembling or to high value-added type industry;
3. Shifts from domestic to internationalization, which implies maintaining a close relationship with foreign countries by promoting the exchange of people, goods, and information.

To reactivate the regional economy, technological revitalization of regional industries must be prompted on a comprehensive basis. To this end, leading-edge technologies must be actively introduced into industry. To keep up with the progress of the information age and international society, the industrial structure, as such, must also be innovated. At the same time, increasingly important roles will be played (1) by research institutes in the technological development that is the basis of new industries and (2) by universities that attract highly educated people who will be engaged in research and development and technical guidance.

The number of high tech businesses in the fields of aviation, space navigation, integrated circuits, computers, optics, industrial robotics, biotechnology, and medical electronics located in the Kinki region is increasing steadily. Such industries established in the region during the period from 1976 to 1983 constituted 7.5 percent of the total number established throughout the country during the same period. These businesses are located mainly along the area's urban axis, connecting Osaka, Kyoto, and Kobe, which has a complete transportation network.

According to the Small and Medium Enterprise Agency's *Report in the Basic Survey of Manufacturing Structure and Activity,* the development of advanced technologies for industry in the Osaka Prefectures is described in terms of industrial ownership including patents (right of new development or invention), utility model rights (right of improvement of existing products to enhance practical utility), and design rights (right of new design, such as shape and color of industrial products). The number of applications for industrial ownership in Osaka prefecture constituted 20.7 percent of the national total in 1977, the ratio gradually decreasing to 19.7 percent in 1980 and to 18.4 percent in 1983. For patents alone, however, the number of applications did not change greatly, as evidenced in the percentages: 17.9 percent in 1977, 17.1 percent in 1980, and 18.2 percent in 1983. In Osaka prefecture, the number of companies with industrial ownership (for self-developed products or technologies) constituted 15.5 percent (26.1 percent for Tokyo) of the total in the country in 1981, while the average number of patents held by an Osaka company in the same year was 109.3 (52.2 for the country, 84.6 for Tokyo prefecture), the highest level in Japan.

As classified by industrial field, the number of industrial ownerships held by companies in Osaka constitutes over 20 percent of the total in Japan in three fields: textiles, steel and nonferrous metals, and transport machinery. For the average number of patents held by an Osaka company, the fields of electrical equipment (702) and general machinery (90.6) have attained very high levels.

THE LOCATION OF INDUSTRY RESEARCH AND DEVELOPMENT INSTITUTES

A study of the locations of private research institutes closely related to high tech industries reveals that the Kinki region accounts for around 21 percent of the national total. These R&D institutes are concentrated along the line connecting the major cities in the Kinki region. Those R&D institutes located around the Osaka/Kyoto axis are mostly related to high tech industries, while the majority of those scattered in other areas are connected with agriculture, forestry, or fisheries.

There are many company research institutes in the Osaka metropolitan area dealing in machinery, textiles, chemical industrial products, electric appliances, pharmaceutical products, and the like. This reflects the characteristics of the industrial structure of the region. A company and its research institutes are not, however, always in the same district, although the proportion of institutes located at or near factory sites is high.

A survey on the factors determining the location of company research institutes in 1981 conducted by the Japan Industrial Location Center found that the places considered most adequate for research institutes were "neighboring prefectures of major cities (Tokyo, Osaka, Nagoya)" (57 percent) and "factory site or vicinity" (30 percent). The location of research institutes is clearly affected by factory locations, but major urban areas are generally preferred despite the governments policy against such concentration.

Reasons for opposing the dispersion of research and development functions mentioned by companies center on information needs such as "difficulties in obtaining market information," "difficulties in obtaining technical information," "difficulties in communicating with headquarters," and the general need for grouping functions, such as "insufficient number of services and industries associated with research and development activities." Many respondents also pointed out the human factor—such as the "unwillingness of researchers and engineers to move" or the "difficulty of recruitment."

Factors closely related to existing companies' locations are also considered important in a 1985 Questionnaire on Location Tendencies of Research Institutions conducted by the Research Institute for Urban Environmental Development. "Distance from the company's factory," "distance from the company's headquarters," and "communication with employees" were among the important conditions that determined the present location of research facilities. The "environment of the neighboring area" was also included as an important factor.

THE LOCATION OF UNIVERSITIES AND HIGH TECH INDUSTRY

High-level education institutions located in the Kinki region total, as of fiscal 1982, ninety-nine universities and 108 junior colleges. These figures comprise 21.8 percent and 20.5 percent, respectively, of the national total (455 universities and 526 junior colleges) and show the historical concentration of universities and other high-level education institutions in this region. Analysis by prefecture shows that universities and junior colleges generally concentrate in the large cities of Osaka, Kyoto, and Hyogo prefectures (82.8 percent and 80.6 percent of the Kinki total).

University campuses are concentrated along the Kyoto/Osaka/Kobe line where railways and road networks overlap. These observations reflect the similar trend seen in the location of high tech industries and research institutes. One characteristic of the high tech industries situated in the OMA is the exceptionally high degree of concentration of electronics industries, including integrated circuits and computers, biotechnology industries centering on medical and pharmaceutical products, and industries associated with new materials such as ceramics and high polymers.

The electronics industry, among the most rapidly growing industries in Japan, is of pivotal importance because it is extensively connected with industries of other fields. Indeed, in Japan it is called "the industry of industries." It certainly is one of the key industries in the future development of the Osaka economy. Electronics-related companies and their research institutes have settled in Osaka since the very early days of the industry. Examples all founded in the 1930s are Matsushita Electric Industrial Co., with 37,000 employees; Sanyo Electric Co., with 20,000 employees; Sharp Corporation, with 23,000 employees; and Minolta Camera Co., with 6,100 employees.

It is said that Osaka has been the center of Chinese medicine imports since ancient times and that Western medical science originated in this city. This historical character serves as the setting for the present concentration of chemical industries in the region. Takeda Chemical Industries, with 11,000 employees; Tanabee Seiyaku Co., with 5,400 employees; Shionogi & Co., with 6,900 employees; Fujisawa Pharmaceutical Co., with 5,800 employees; and other major chemical manufacturers located in Osaka between 1918 and 1934.

The cotton spinning industry that prospered at the end of the nineteenth century earned Osaka the nickname Manchester of the Orient. The great concentration of textile factories continues even now; among the major manufacturers are Unitika, Ltd., with 5,700 employees; Shikibo, Ltd., with 2,800 employees; Gunze, Ltd., with 4,800 employees; and Kanebo, Ltd., with 6,200 employees. These companies are currently advancing into the related fields of new materials, including high polymers or composite materials, and biotechnology. Recent years in the OMA have seen the settlement of major Japanese companies dealing with new ceramic materials—namely, Murata Manufacturing Co., with 2,400 employees, and Kyocera Corp., with 13,000 employees, both founded in the 1950s.

The factories and their research institutes and the universities and their research institutes are not necessarily found in the same areas. This is thought to be attributable to the selection of new sites on the basis of such factors as the time and distance from existing facilities (headquarters, factories, head offices of universities, and so forth), and the availability of land. In any case, extensive and varied exchange between university researchers and researchers of enterprises is *the* decisive factor when companies try to advance into a new field or consider establishing a new research institute. This is reflected in the enthusiasm that enterprises show when they scout excellent manpower from universities and other high tech companies and when they send their own researchers to universities and other research institutes for an extended period of time.

For this reason, areas around major universities are increasingly involved in urbanization programs, rendering land difficult to obtain and expensive. The result of this situation is the concentration of newly established high tech factories and research institutes on the line connecting the three major cities, where communication with universities is easier and travel is more convenient.

NEW TECHNOLOGY-BASED SMALL BUSINESS

Small businesses occupy the main position in Japan's national economy in that, as of 1983, they include 99.5 percent of the total number of business establishments, 73.9 percent of the employees, and 52.4 percent of the product shipment value. In particular, this group plays a central role in the OMA with 99.8 percent of the business establishments, 83.6 percent of the employees, and 71.9 percent of the product shipment value for Osaka City in 1983. It is virtually impossible to think of upgrading the regional economy without promoting small business.

Many of these companies are involved in research and development activities and the development of new products. Various companies are cooperating with U.S. firms in the fields of hardware and software. Local governments have instituted diverse systems that offer public financing for small business activities and in this way are trying to facilitate the introduction of more advanced technology into these enterprises. Some of the financing systems offered in Osaka are listed in Table 2-2.

The Osaka Science and Technology Center is utilized as a place of interchange between industry, government, and academic institutions. In the 1950s local businesspeople and public administrators realized that there was a need to promote science and technology, to improve deficiencies in scientific and

Table 2-2. Financing Systems for New Technology Businesses.

Financing System	Sponsor	Remarks
Loan for new-technology businesses	State-owned Smaller Business Finance Corporation (SBFC)	Loan to 1 company in approximately ¥160 million (1985) $1.2 million
Loan for leading-edge technologies	SBFC	19 companies, approximately ¥1,770 million (1985) $13.1 million
Loan for information infrastructure	SBFC	9 companies, approximately ¥160 million (1985) $1.2 million
Loan for leading-edge technological R&D	Osaka prefecture	17 companies, approximately ¥490 million (1985) $3.6 million
Loan to develop R&D-oriented companies		30 companies, approximately ¥970 million (1985) $7.2 million
Financial guarantee for investment in mechatronic equipment	Osaka prefecture	52 companies, approximately ¥1 billion (1985) $7.4 billion
Financial guarantee to smaller businesses for leading-edge technological development expenses	Osaka City	Established in 1983. 12 companies, approximately ¥260 million (1985) $1.9 million
High tech loan		Established in 1986. 12 companies, approximately ¥500 million (1986) $3.7 million
Obligation guarantee for R&D expenses	Venture Enterprise Center	
Subscription of additional stocks	Osaka Small Business Investment Company, Ltd.	

Source: Comprehensive Planning Bureau, Osaka Municipal Government, 1987.

technological research, and to develop a system of mutual cooperation between public and private research institutions to eventually bring about the recovery of Osaka's economy. Based on this understanding, the Osaka Science and Technology Center was established in 1960 (its name and organization changed in 1967). The center is located in Osaka and is supported by contributions from local industrial associations with financial assistance from the Osaka prefectural and municipal governments and the Science and Technology Agency.

The center's exhibition rooms, conference rooms, training facilities, and salons serve a series of functions, including as a think-tank for surveys on research and technical development and for gathering information from industrial, government, and academic worlds. The center is also involved in promoting technological development of leading small-size enterprises as well as the diffusion of scientific and technical knowledge and information to the public.

A recent example of interchange between industry, the public sector, and the academic world was the cooperation in technical development in the fields of electronics and new materials which began in 1982. Four manufacturers of electric home appliances in Osaka (Matsushita Electric Industrial Co., Sanyo Electric Co., Sharp Corp., and Mitsubishi Electric Corp.) joined, with the support of the Osaka Bureau of International Trade and Industry, in developing a "new microprocessor for electrical home appliances." Research and development, conducted with cooperation of Osaka University, successfully resulted in software and chip production. Similar interchanges beween industry, the public sector, and academic institutions are under way in the field of new materials technology, where manufacturers and universities are combining their effort in research on shape-memory alloy and high polymer materials for medical use.

It is clear that the most important role to be played by the Osaka Science and Technology Center is for it to serve as venue for coordinating manpower and information exchange. It is also worthy of note that this facility has been set up at the very core of a central city in the metropolitan area.

TECHNOPOLIS OSAKA, FULLY INTEGRATING
EXISTING URBAN FUNCTIONS

Conventional industrial structure is undergoing significant transformations that generate such rapid economic and social change that successful high tech enterprises of yesterday are found to be declining today. In Japan, industry is rapidly restructuring itself by the process of "scrap and build." The crucial factor for pioneering enterprises of the new era to cope with drastic socio-economic change is to develop versatile minds capable of developing new technopoleis and carrying out creative research. The role played by an R&D center that enables free interchange among industrial, governmental, and academic sectors will be important in this context.

Creativity is indispensable for research and the development of advanced technology. Many specialists claim that while creativity is enhanced by exchange of information, face-to-face contact is far more efficient than media services or electronic exchanges. This belief encourages the development of various places and opportunities for researchers to be able to freely communicate with each other. Thus far, such places and opportunities have been provided in the central cities of metropolitan areas. Such cities are network hubs that furnish diverse facilities and amenities. However, these central cities need to further improve their function as facilities of face-to-face meeting places for researchers of the future.

Accordingly, it might be appropriate to imagine the optimal technopolis within a metropolitan area as one that fully integrates existing urban functions while taking advantage of existing facilities and ensuring easy access. Based on such assets, a knowledge-intensive and high-value-added R&D center should be constructed only a short distance from the city. Face-to-face meeting places for researchers should also be provided within the city. Furthermore, land prices should ideally be relatively cheap to permit acquisition of a sufficiently large area for the R&D center.

In summary, the key point for developing Osaka into a technopolis that is fully integrated with existing urban functions is to attract creative persons by providing an inviting, attractive urban environment and by developing traffic and communications networks to link the various places within the region. To this end, the OMA is embarking on two primary projects that will be the driving force for restructuring the area as well as for developing the regional economy for the coming internationalized information society: (1) construction of the Kansai International Airport and (2) building the Kansai Science City, where researchers will assemble for high tech development and creative studies in a joint effort with enterprises, the public sector, and academic institutions.

These two leading projects will provide the major infrastructure necessary for the technopolis of the OMA for the twenty-first century. The formation of a technopolis that fully integrates existing urban functions, however, will require other features. Although Osaka has been growing as the center city of the OMA, it needs to improve those urban functions that will be suitable for the coming information society and that will be attractive to creative researchers—namely, advanced technology. In addition, an alluring urban environment will have to be prepared with good housing and living conditions; a highly developed medical system; a favorable educational environment and opportunities; cultural, artistic, sports, and recreation facilities; and open-minded local communities. Such qualities will ensure the gathering of creative persons and the facilitation of face-to-face interchanges that are key factors in the creation of a successful technopolis. Based on this principal strategy, the city of Osaka intends to promote two of its own projects so that the technopolis can fully integrate existing urban functions in the coastal zone: the Naniwa Necklace Project and the Technoport Osaka Project (see Figure 2–2).

Naniwa Necklace Project

In Japan it is said that creative researchers involved in high tech research and development are strongly attracted to large cities. The reason for this is the variety of activities provided by the urban environment, which serve as background and support for such creativity. Large cities have the power to make life exciting, charming, and enriched.

The central area of Osaka, capital of Japan as early as the seventh century, was urbanized at the end of the sixteenth century and developed into a national economic center (known by the nickname *tenka no Daidokoro*, meaning "the major emporium of all goods and merchandise") between the seventeenth and nineteenth centuries.

Today Osaka has a northern railroad terminal that is transited daily by some 2.3 million passengers and a southern terminal transited by about 1.15 million passengers each day. Both terminals have a shopping-entertainment zone where department stores, restaurants, and other facilities are concentrated. A central business district spreads between the northern and southern terminals, with streets lined by public offices, finance, and insurance companies as well as other enterprises and headquarters. It is the hub of the OMA's economy, where more than a million persons work each day.

A railway line, in the form of a loop, surrounds the central area and along this railway loop a series of projects are distributed that are intended to redevelop former railway cargo yards and factory sites, roughly in the shape of a necklace around the central area of Osaka. By promoting these projects, Osaka expects to revitalize its central area while at the same time creating spin-offs for the surrounding zones. This effort is called the Naniwa Necklace Project. *Naniwa* is the ancient name for Osaka. The title refers to the layout

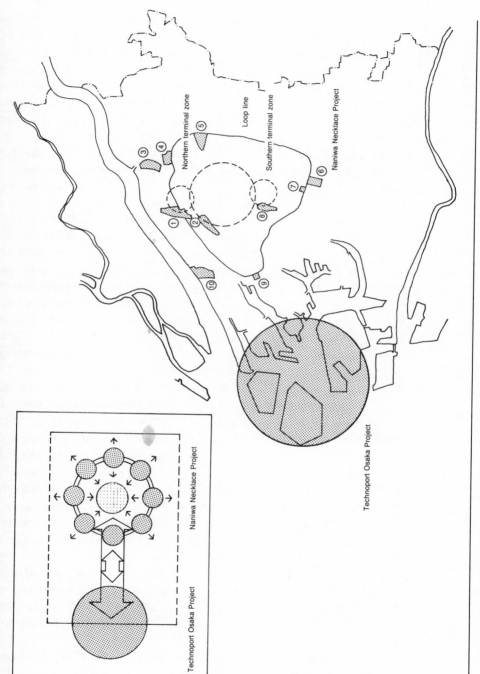

Source: Comprehensive Planning Bureau, Osaka Municipal Government, 1987.

of the various redevelopment projects that will work to trigger the renovation of old office buildings of the central area into new intelligent buildings; the construction of an international exchange center and good-quality urban housing and culture and art centers; the improvement of city amenities such as the development of the waterfront and open space networks; and the renovation of the traffic infrastructure. The Naniwa Necklace Project presents a new concept of redevelopment, with the objective of fostering an attractive urban environment in the city and increasing its drawing power.

Some of the redevelopment programs included in the Naniwa Necklace Project have already been implemented. The major ones are detailed below (also see Table 2–3):

1. The creation of a new city base is planned around an area formerly used as a railway cargo yard. This is the only large unused land area left in the shopping-entertainment zone of the northern terminal, which is otherwise crowded with department stores, restaurants, and other high-rise buildings. A futuristic terminal-type complex will be built on the former cargo station site to the north of Osaka Station, the base for Japanese National Railways (JNR) operations in Western Japan. There will be facilities to extend various services related to business, information, culture, and international matters to the western half of Japan, as well as parks and cultural centers to be used by the residents of the area.

2. Another new project will soon be constructed on land formerly used as JNR cargo station. The project in the Minatomachi area will create a complex that in tandem with the City Air Terminal will connect the Kansai International Airport with the city center and will serve an international exchange function. It will also be in an area closely associated with the fashion industry.

3. To the east of the central business district, a new business precinct of the twenty-first century is emerging on former manufacturing sites. Called Osaka Business Park (OBP), it is gradually forming into a futuristic office center. Adjacent to it, Osaka Castle Park offers parks and extensive open space that is difficult to find elsewhere within the city. In this setting of rivers and greenery, skyscrapers and smaller edifices will be constructed, following the superblock system, without being in too close a proximity to each other. The City of Osaka will thus be provided with new types of business, commercial, and service activities along with cultural and information facilities.

   Some buildings have already been completed in the OBP, and more are now under construction. The superhigh Twin Building, for example, is an "intelligent" building 150 m high constructed by a real estate firm of the Matsushita Electric Group. Fujitsu, Ltd., among the top corporations in the computer and communications field in Japan, has moved into another intelligent building that is expected to become a base for their operations in Western Japan. NEC Corporation will also be moving into an intelligent building now under construction, with similar facilities.

4. One of Japan's largest redevelopment projects is being implemented by the municipal government in Abeno, at the south end of the central area. The plan's objective is to construct an attractive city subcenter where commercial and business establishments and urban residences coexist without interfering with each other. An area of 18 ha (69.2 acres) that is now crowded with houses and is adjacent to Abeno Terminal (the third largest in Osaka) will be redeveloped.

5. Nakanoshima Island has long been a favorite place of the area's citizens. Located in the CBD, the island, around 55 ha (135.9 acres) in the Okawa River, has a park and a beautiful waterfront but also is a center of administration, business, and culture. The City Hall, the Central Public Hall, the Osaka branch of the Bank of Japan, and numerous other company headquarters are located here. The western half of Nakanoshima, however, is a less developed area of about 30 ha (74.1 acres) with warehouses and other facilities.

A plan is now under study to create a prettier urban landscape in the western area of Nakanoshima, while preserving its waterfront and greenery. The objective is to develop the infrastructure needed to run the area as a base of international exchange and cultural activity. Construction of a science and technology museum and a modern arts museum is begining along with plans for an international conference hall and other facilities.

6. The city is also urbanizing the Yodogawa riverside area, which is conveniently located about 2 km (1.2 mi) to the northeast of the northern terminal. The urbanization of this zone aims at creating a fine residential environment on the waterfront of the Okawa River with parks and good-quality housing units located near the residents' workplace.

An area of 35.6 ha (87.9 acres) faces the river. Formerly there were several factories here, but under the ongoing program the area will be transformed into a comfortable, convenient, and attractive urban zone with a park facing the river, elementary and junior high schools, and urban housing providing approximately 3,200 dwellings.

7. Similar plans comprising the integrated redevelopment of urban residential areas are being implemented on former factory sites in Takami (51.8 ha or 127.9 acres) and Sakuranomiya Nakano (29.8 ha or 73.6 acres) areas as part of the Naniwa Necklace Project.

Table 2-3. The Naniwa Necklace Project.

No.	District	Project Outline	Location, Area	Major Facilities Planned
1	Umeda	Development into an international center of information processing and transmission, by taking advantage, for commuting purposes, of this district's adjacence to Japanese National Railways' Osaka Station Strengthening the terminal function as well as the cultural/educational function through construction of culture centers etc.	Northern terminal area Former site of freight yard, south of Osaka Station: 9.2 ha; 22.7 acres freight yard, north of Osaka Station: 26 ha; 64.2 acres total: 35.2 ha; 86.9 acres	New railroad station, urban expressway CBD park, office buildings, culture center, press complex
2	Minatomachi	Integrating advanced commercial functions into a fashion hub of the 21st century by making full use of the existing concentration of fashion-related commercial functions and traditional culture in the southern terminal Upgrading the international exchange function by constructing hotels and other facilities around the City Air Terminal, which will link the Kansai International Airport and Osaka's central business district	Southern terminal area. Around the former site of the JNR freight station: 17.5 ha; 43.2 acres	New railroad station, urban expressway City Air Terminal, office buildings, hotels

3	Osaka Business Park	Development into a "town in the park," adjacent to Osaka Castle Park and rich in greenery and open space Future-oriented development by providing new commercial/business/services and cultural/information functions	Former site of factories between rivers east of the CBD: 26 ha; 64.2 acres	"Intelligent" office buildings, hotels, promenade and new railroad station, using the superblock approach
4	Abeno Urban Renewal	Redeveloping this dilapidated and overcrowded housing district into an urban subcenter supported by the railroad terminal, essential to developing southern Osaka	Urbanized area near Abeno terminal, southernmost of the central area: 28 ha; 69.2 acres	Office buildings, hotels, gymnasia, medical facilities, cultural facilities, urban housing, parks
5	Western Nakanoshima	Development into an international and cultural exchange by taking advantage of the comfortable environment rich in water and greenery	Western Nakanoshima in Okawa River: 30 ha; 74.1 acres	Science and technology museum modern art museum
6	Yodogawa Riverside	Developing a comfortable, convenient environment by taking advantage of this district's qualities as a verdant waterfront and by constructing good-quality urban housing (in proximity to workplaces), parks, schools, and other facilities	Former site of factories along the Okawa River: 35.6 ha; 87.9 acres	Housing (3,230 units), parks, junior high school, walkways

Table 2–3 continued.

No.	District	Project Outline	Location, Area	Major Facilities Planned
7	Takami	Comprehensively constructing good-quality urban housing (in proximity to workplaces), parks, schools and other facilities and thereby developing an environment rich in flowers and greenery	Former site of factories along the Yodogawa River: 51.8 ha; 127.9 acres	Housing (approximately 3,400 units), elementary schools, neighborhood parks
	Sakuranomiya-Nakano	Constructing good-quality urban housing, parks and other facilities by taking advantage of an environment rich in water and greenery Utilizing the urban development ability of the private sector in facility construction	Former site of the JNR freight station near the Okawa River: 29.8 ha; 73.6 acres	Bus terminal, high-class housing, gymnasia, culture centers, shops, etc.
8	Benten-cho	Development into an urban subcenter to serve as an amenity core with business/commercial facilities Urban planning fully utilizing the urban development ability of the private sector	Municipal land west of the central area: 3 ha; 7.4 acres	Design competition is underway
	Kasumi-cho	Introducing attractive entertainment and commercial/business facilities to revitalize this district, embracing Shinsekai, a longstanding amusement quarter	Former site of municipal streetcar depot in the southern central area: 2.5 ha; 6.2 acres	In planning

Source: Comprehensive Planning Bureau, Osaka Municipal Government, 1987.

8. Other development projects are also under way in the Benten-cho (13 ha or 32.1 acres) and Kasumi-cho (2.5 ha or 6.2 acres) areas.

Two additional undertakings, which are being implemented concurrently with the Naniwa Necklace Project, aim at strengthening the information and communications network of the entire region. These are being promoted by the city of Osaka and take advantage of the liberalization of telecommunication operations stipulated in the new revised law of April 1985:

1. *Strengthening the regional communications network.* If communication services are offered by multiple enterprises, this will eventually bring about a reinforced and better information infrastructure, including greater safety for cities in regard to information. Moreover, the cities themselves will be revitalized, as diversified needs of users are met. With a view to these objectives, in October 1985 the city of Osaka established the Osaka Media Port Co. (OMP) in cooperation with the private sector. The OMP will lay optical fiber cables utilizing existing municipal subway and expressway networks, thus building a regional digital communications network that will offer services in addition to the system set up by the Nippon Telegraph and Telephone Co. (NTT). The project, presently in its first phase, is developing a network that will serve the city of Osaka and nine neighboring municipalities. The service area will eventually be expanded to cover all of the Kinki region, making use of the electric power supply cable network of the Kansai Electric Power Co.

2. *Project for commercializing a video information system.* This sophisticated data processing system will permit integrated operation and processing of images, characters, voice, and other forms of information transmitted by computers—namely, video-aided networks (VAN), videotex, computer-aided television (CATV), and computer-aided design (CAD), which have thus far been treated separately. The system, when introduced in the CBD, especially in areas where apparel and related industries concentrate, will contribute to the revitalization of these areas by offering diversified technical information and data bank services useful to management. Such services might include broadcasting of fashion shows and other events, the transmission of product catalogues by video mail, and design support for CAD users. This commercialization project was inaugurated in May 1986 and is jointly sponsored by the Japanese government, the Osaka municipal government, and private companies. Surveys and research are currently being conducted on the development of terminals, other hardware, and the type of information to be provided.

In addition to the projects mentioned above, Osaka is endeavoring to create a beautiful urban area by constructing waterfront parks (Osaka was once known as the City on Water), improving the urban landscape, expanding open spaces by developing pedestrian promenades and urban plazas, transforming urban street intersections into small parks, and planting trees and flowers in divider strips. In addition, the city is undertaking to improve its amenities and increase its attractiveness by such action as proposing guidelines in regard to building appearance and by lighting nighttime landmarks.

Technoport Osaka Project

Since ancient times Osaka City has been expanding its coastal areas by reclaiming the sea to obtain new space for further urban development. Currently in Osaka Bay, construction of three man-made islands is underway, the expected area of these islands will total 1,545 ha (3,816.2 acres). The Nanko district (930 ha or 2,297 acres) has already been completed on one of the islands. The Hokko north district (225 ha or 555.8 acres) island is recently finished. The Hokko south district (390 ha or 963.3 acres) is scheduled for completion in the late 1990s.

These three islands will be connected to the center of Osaka City and to other major cities in the OMA via railway and road networks, which will be further extended to the Kansai International Airport currently under construction. To rapidly expand the urban functions of Osaka City, the Technoport Osaka project is being promoted for the three islands. The name *Technoport* was created by combining the prefix *techno* from technology, a key for future development, with *port,* gateway for the flow of people, goods, and information. The project is aimed (1) at creating a future-oriented town along Osaka Bay that can meet the needs of the information society and (2) at constructing a new international port of exchange built on the most updated information, knowledge, advanced technologies, and highly educated people.

These three man-made islands will be zoned into areas of different functions: highly sophisticated communication, international trading, and development of leading-edge technologies. Each area will be provided with advanced facilities. Other functions will also be added for sporting and other recreational activities and a comfortable urban residence, thus creating a future-oriented city comprehensively serving various human activities. In the Nanko district, where land utilization has already commenced, the Technoport Osaka Project is being actively promoted.

First, the Highly Sophisticated Communication Area will serve as the communication center of Technoport Osaka. A teleport will be constructed to establish a satellite communications network that will permit Osaka to exchange information directly with many cities throughout the world. The construction of optical fiber networks is also being planned for communication with the center of Osaka City and with other cities in the OMA. This area will also contain clusters of corporations specializing in the development of information-processing-related software and VAN data banks, and the information-processing departments of other companies. Construction of a satellite antenna has just begun in the Nanko district.

Second, the World Trade Area will have three functions: world trade and transaction, international exchange, and comprehensive distribution for domestic and overseas cargo transportation. The International Exhibition Center, Osaka (INTEX OSAKA), completed in March 1985 in the Nanko district to fulfill the need for world transactions, comprises INTEX Plaza, with a new multipurpose hall available for various events. With a total area of 13 ha (32 acres) and an exhibition area of 4.5 ha (11.1 acres), INTEX OSKA, a world-class advanced exhibition hall, provides, in addition to various trade fair facilities, a video data bank for storing information on companies and commodities, and INTEX INFOR, which furnishes information required by visitors on events and their locations, transportation, and sightseeing trip information.

In the vicinity of INTEX OSAKA will be located trading-related businesses and the World Trade Center, comprising hotels and trading companies, thus creating a complex of world transaction, trade, and international exchange functions. For comprehensive distribution, this area is also planned to serve as a key point of land, sea, and air cargo transportation. Also planned for this area is construction of the Air Cargo City Terminal (ACCT), which will function as part of airport customs to enhance the convenience of air transportation between cities. It will include foreign trade container facilities to keep up with increased trading volume and a comprehensive distribution-related information center.

Third, Technoport, a future-oriented city of the twenty-first century, will be provided with a range of sporting and other recreational facilities and residential functions, thus furnishing a comfortable urban space for citizens. At the Hokko north district, other amenities are also planned such as a large recreational and sporting area (100 ha or 247 acres) and Hokko Marina, a marine sporting base in Osaka Bay.

At the Nanko district, Nanko Port Town is already completed and has an expected population of 40,000. The port has educational facilities, CATV for individual households and the world's first large-scale regional pneumatic municipal-waste conveying system. Despite its being a new town constructed on a man-made island, Port Town offers space for communing with nature by banning automobiles and creating rivers and greenery. At the northwest and southwest ends of Nanko island are located a natural bird sanctuary (19.3 ha or 47.7 acres) and a fishing park (300 m or 328 ft long), the adjacent area being provided with a unique man-made swimming beach (3 ha or 7.4 acres) for Osaka citizens. Port Town is connected with the center of Osaka City by a new train and subways.

Finally, the High Tech Lab Area is planned to provide space for research in applied science and production technology crucial for companies in the advanced technology development race. This area will be directly connected by expressway with Kansai International Airport and Kansai Science City, to permit face-to-face communication among domestic and overseas researchers. Construction of governmental and private-sector facilities for research in basic and applied science is currently under planning in this area. It is intended to serve as a center of technological development in the fields of telecommunications, mechatronics, biotechnology, fermentology, new materials, energy, and ocean engineering.

Also planned for this area is the construction of a computer center open to the public, a technological and patent information center, a library, and various facilities for the training and the communication among researchers. A research park and an industrial park will also be created in this zone as a basis for nurturing venture businesses in microelectronics, optoelectronics, medical electronics, mechatronics, new materials, and biotechnology. For newly founded businesses and small and medium enterprises, other projects are also planned, including the establishment of technical and managerial consultation and financing systems, so that the entire area will serve as an incubator for such companies.

In Nanko district, System House was completed in summer 1987. System House is made up of leading small and medium-size enterprises that use advanced applied technologies in microelectronics, software development, and mechantronics. FANUC, a leading company in the field of robotics, opened the Kansai Technical Center in March 1986. Mitutoyo Manufacturing Co., specialists in precision machinery, is also located in this area.

The Osaka Technoport Project on the three man-made islands will serve as a key to transforming Osaka into a technopolis integrating existing urban functions. This technopolis is aimed at the total harmony of technology, human beings, and nature in twenty-first century society.

CONCLUSION

The OMA, boasting a history of more than 1,500 years, has achieved an impressive concentration of urban functions through the synergistic development of closely located core cities, including an economic center in Osaka, cultural and art centers in Kyoto and Nara, and the growing trade center in Kobe. As Japan enters the era of internationalization and information, Osaka is preparing to restructure its local communities and promote local development. To this end, two leading projects are underway: the Kansai International Airport, to serve as an international gateway for people, goods, and information; and Kansai Science City, to serve as a hub of research and development.

To construct a new urban technopolis by making full use of the existing concentration of urban functions, the OMA must invite creative persons for innovative technological development and creative research. OMA must also provide them with opportunities for face-to-face exchange. Being fully aware of these requirements, the OMA is implementing the Naniwa Necklace Project, which aims at the revitalization of old urban districts, and the Osaka Technoport Project, which encourages the placement of leading-edge businesses on the coastal area to help develop Osaka into a futuristic technopolis. Through comprehensive implementation of these and other projects, Osaka is committed to developing into a technopolis that fully integrates existing urban functions.

Chapter 3

TSUKUBA SCIENCE CITY COMPLEX AND THE JAPANESE TECHNOPOLIS STRATEGY

Masahiko Onda

Japanese successes in such industrial fields as textiles, steel, automobiles, and electronic appliances were not attained merely by genius, imagination, or creativity. More accurate clues may lie in the teamwork on production lines, hard-working employees, and culturally conformed deliberate habits of simple labor, all of which are sociocultural indicators. In addition, it is said that Japanese R&D efforts are mostly made along manufacturing lines by in-situ engineers. If this is true, technology development in and of itself does not mean much. Because of the transferable nature of technologies, they can be bought or traced in most circumstances. Nevertheless, the Japanese were driven to produce basic technologies because other nations' markets have been somehow diminished by Japanese commercial success.

The Japanese government was not entirely enthusiastic about enacting strong promotion policies for basic sciences or technologies. R&D administrators' attitudes, however, particularly in the Ministry of International Trade and Industry (MITI), have changed. Certain officials once referred to *kiso* (which in Japanese means "basic research") as being little more than *kuso,* which means "crap," and implied that MITI was essentially a commerce-oriented ministry. It was felt that MITI could lead the country to industrial success through trade. This confidence in MITI has been transformed, by

This chapter does not reflect official views of the ministry to which the author belongs or the policy of the Japanese government.

necessity, to an awareness that Japan must find another way to develop its role in the world economy. In this context, basic research in science and technology has become more important, especially in response to the recent intense trade frictions that Japan is now facing.

This chapter distinguishes between conglomerate facilities of product manufacturing and those of an intellectual property production nature. The term *research park* is used to indicate an aggregate group of research-oriented facilities with a certain level of integrity. The term *industrial park* is used for an aggregation of mass-production type industries.

TSUKUBA SCIENCE CITY: THE STATUS QUO

Tsukuba was planned and developed by the public sector, and its clearly delineated zones are allocated to various specific uses. The following sections present national and prefectural government plans related to Tsukuba.

Location

Tsukuba Science City is located thirty-five miles northeast of Tokyo and twenty-five miles northwest of Narita International Airport. It consists of five already existing towns and one village, all of which are administratively independent. The science city's total surface area amounts to 28,560 hectares (11,563 acres). At the center of the city, there is what is called a research and educational district of 2,700 hectares (1,093 acres). This district covers a strip of land eighteen kilometers (eleven miles) north/south by six kilometers (four miles) east/west. Over half of this academic district is held by forty-six national research and educational agencies and a few private research institutes. These national research institutes are involved in various disciplines, representing most government scientific, technological, and social issues covered by ten ministries and agencies. This district has 100 parks of various sizes, which amount to 100 hectares (forty acres) and take up 4 percent of the area. About a quarter of the area is occupied by residential and commercial sectors. This district is served by a well-developed network of pedestrian and bicycle pathways, with a total length of fifty kilometers (thirty-one miles) connecting all sections of the academic district.

Public and Private Research Parks

These forty-six national research agencies compose 30 percent of all Japanese national research agencies and 40 percent of all personnel. They expend 50 percent of the total R&D budget. As shown in Figure 3–1, the rest of the Tsukuba Science City area is a suburban or development district, where there are three major research parks for private industries: Northern Research Park, Western Research Park, and Toukoudai Research Park. Toukoudai Research Park was constructed in 1982 by the landowners' association with the help of the Japan Housing Corporation (Hihon Jutaku Koudan) under the Ministry

Figure 3–1.   Tsukuba Research and Industrial Parks.

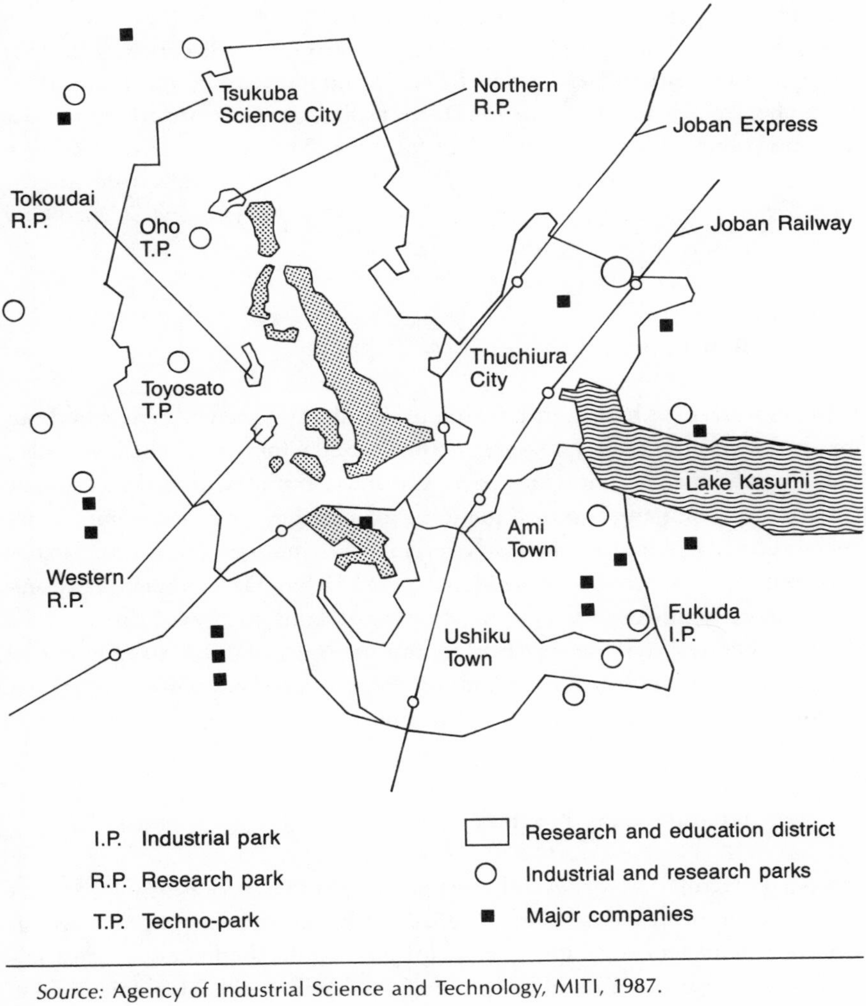

Source: Agency of Industrial Science and Technology, MITI, 1987.

of Construction. It has forty hectares (sixteen acres) on the west side of the science city.

At present, about thirty private enterprises have located their high tech R&D laboratories under the name of Tsukuba Consortium. In this park, seven ERATO (Exploratory Research of Advanced Technology) projects have been in progress from 1981 under the sponsorship of the Research Development Corporation, an affiliate of the Science and Technology Agency. The annual funding level of ERATO is Y9 billion ($64.3 million, calculated at Y140 to the dollar). ERATO covers subjects ranging from biosciences to solid physics, to microtechnology regarding semiconductors. In Northern Research Park, fourteen companies are now planning to install research facilities over a seventy-hectare (twenty-eight-acre) area. Western Research Park has fifty-six hectares (twenty-three acres), and fifteen companies are going to construct research labs. Within the Academic District, there are three private companies that are trying to set out their R&D facilities. The Northern and Western Research Park areas are under preparation by the Ibaraki prefectural government. A clear planning policy is that the core of the Tsukuba Science City should not accept mass-production factories.

Industrial Parks

Besides these three high tech research parks, Ibaraki prefecture has laid out one industrial park for mass-production type factory use, which is called Fukuda Industrial Park. It is located in Ami Town southeast of Tsukuba Science City. Sixteen companies are now planning to open their factories over its eighty hectares (thirty-two acres) territory. Several other industrial parks for factory use are under preparation by the Housing and Urban Development Corporation (Jutaku-Seibi-Toshi-Kodan), a Ministry of Construction's affiliate. As is shown in Figure 3–1, these are called techno-parks. In addition to these, several prestigious companies including trading companies and newspapers have their liaison offices in Tsukuba to collect technical information.

Technolinkage Program

Ibaraki prefecture has for several years proposed the idea of forming contacts among neighboring cities, which is called the Technolinkage Program. About sixty-five kilometers (forty miles) from Tokyo, there are a number of industrial-commercial areas that are still in the process of development. Therefore, if these cities are organically linked by efficient transportation and communication, they will form a part of the greater metropolitan area. In this sense, Tsukuba Science City could play a go-between role for these industrially important areas. Under this concept, Tsukuba is to be linked with Narita International Airport and Utsunomiya Technopolis and preferably also with cities in the Tohoku district. This idea is shown in Figure 3–2. Tsukuba Science City and its peripheral industrial areas are planned to be connected to existing neighboring industrial cities like Mito, Hitachi, and Kashima industrial areas as well as Narita International Airport and Tokyo, which are located within a forty-mile radius.

HISTORY AND GROWTH OF TSUKUBA

Tsukuba was first conceived in the 1958 metropolitan area development plan, in which Tsukuba was envisioned as a satellite science city of Tokyo. This plan was a modification of the Greater London Area Plan consisting of existing urban areas, green belts, and outer areas to be developed. The second metropolitan area reform plan, presented in 1968, was similar to the first concept but with a modified definition of green belts. The third plan emerged in 1976. In this plan, the greater metropolitan area was defined as a center city with a number of urban areas to be developed around the outskirts. These outer urban areas were not to be simply satellite cities but to have both residential functions and industries. Therefore, these outer sites are planned as independent and self-sustaining cities. This idea continues to have support.

Apart from these metropolitan areas reform plans, a new science city layout was produced. As shown in Table 3–1, construction of Tsukuba Science City was officially started in 1970 between the second and third metropolitan reform plans by the enactment of the Tsukuba Science City Construction Law. In 1980 relocation of forty-three national research institutes and universities was completed at a cost of almost Y2 trillion ($6.5 billion). In 1985 Tsukuba Science Exposition attracted over 20 million visitors from all over the world.

Population Growth and Its Balance

Tsukuba Science City is still administratively divided into six municipal bodies whose total evening population in January 1987 was 152,000. This figure, however, is far below the planned population of 220,000. By the year 2000 the population of the Tsukuba area including the neighboring cities of Tsuchiura and Ushiku is expected to reach 700,000, double the present population of 350,000.

Figure 3–2.   Techno-Linkage Concept.

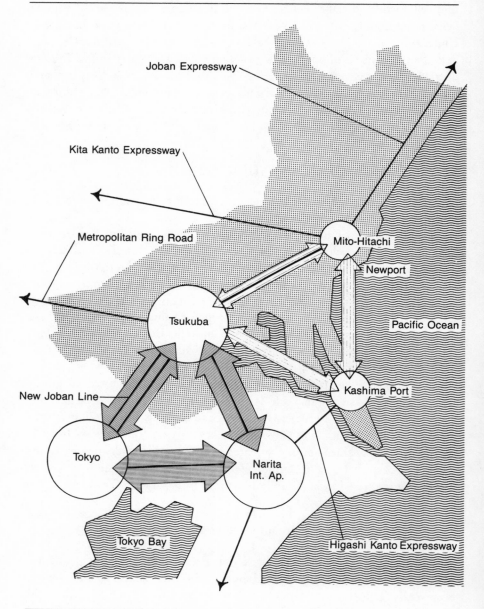

Source: Agency of Industrial Science and Technology, MITI, 1987.

Table 3-1. History of the Tsukuba Science City
Construction Project.

Year	Event
1961	Prime Minister Ikeda's cabinet inquiry into possible relocation of government agencies.
1962	The Council for Science and Technology proposal for relocation of national research agencies.
1963	Cabinet approval of the Tsukuba area for a new science city. The Japan Housing Corporation (later, the Housing and Urban Development Corporation) assigned for land acquisition and preparation.
1964	Cabinet decision for ten-year construction from 1965 with target for Tokyo Olympic Games.
1968	Construction started for the National Research Center for Disaster Prevention.
1970	Legislation for construction of Tsukuba Science City promulgated.
1972	The National Research Institute for Inorganic Materials opened as the first agency to settle in Tsukuba. Cabinet selection of forty-two research and educational institutes (forty-three in 1973) for relocation.
1973	The University of Tsukuba (the former Tokyo Educational College) founded. Oil shock retards planning.
1974	The National Land Agency (*Kokudo-chou*) established.
1975	Cabinet decision to postpone target date for completion to 1979.
1980	Forty-three research and educational institutes completed and started operations.
1982	Mitterrand and Thatcher visit Tsukuba.
1985	Tsukuba Science Expo '85 held. Over 20 million participants.

What kind of people are the most representative residents in Tsukuba Science City? They can be divided into four groups: (1) government-employed researchers working at national institutes and educational facilities (this population includes 35,000 households, including local civil servants); (2) commercial people engaged in small businesses; (3) farmers who have been living in the area since long before Tsukuba's designation as a science city (residents who belong to groups 2 and 3 total 90,000 to 100,000); (4) approximately 10,000 university students.

Local industrial demographic statistics reveal professional population distribution and change in the 1970 through 1980 period. The primary industry-related population decreased between 1970 and 1980 from 57.3 percent to 26.2 percent. The population of secondary industry increased slightly from 16.6 percent to 21.7 percent in the same period. The tertiary industrial population increased significantly from 26.1 percent to 52.1 percent. This data demonstrates that the Tsukuba Science City area is becoming an increasingly service-oriented region. The tendency after 1980 will deviate slightly from the previous period since the secondary industry population growth seems remarkable. If Tsukuba Science City were to be a self-contained city, then some 70,000 resident job opportunities would have to be created to strengthen the commercial sector, which would eventually support the R&D sector.

Tokyo's Population Explosion and Tsukuba's Role

In the Tokyo suburban area, the population is anticipated to grow by 2 to 2.8 million by the year 2000. However, Tokyo's bedtowns or peripheral cities like Ohmiya, Urawa, Tachikawa, Chiba, and Yokohama will be unable to absorb this level of population growth. The Tsukuba area offers relatively inexpensive land and provides the last remaining area near Tokyo left undeveloped. The Tsukuba area is expected to absorb some of this population growth, although the amount will be relatively modest. Tsukuba is also supposed to play an important role as a shelter in the event of a disaster in the metropolitan area.

Why has the Tsukuba area been left untouched despite its proximity to Tokyo? Why has the economy of the Tsukuba area been less developed compared to other peripheral areas of Tokyo? One major reason is the infertile soil of the region, rendering low agricultural productivity. Another reason is the social and historical background of the region, which has been ruled by many petty feudal lords.

Tsukuba Science City is presently considered an isolated island, remote from normal human society—a city of academics and education. Tokyo, on the other hand, presents a disastrously overcrowded living situation. For an average Japanese employee, it is a life-long goal to acquire his own house at a reasonably close distance from his work. Despite the main theme of Tsukuba Science Expo, "pursuing an ideal living space," the average Japanese feels hopelessly desperate about the contemporary Japanese housing problem. This situation is not expected to improve in the near future. Still, some hope remains to have Tsukuba contribute to the relief of Tokyo's congestion through the National Land Agency's concept of reforming the greater Tokyo area into a cluster of independent cities.

TSUKUBA'S INTERNAL PROBLEMS AND ITS FUTURE

Newcomers versus Old-Timers

Tsukuba has begun to mature. Although new residents have settled in fairly well, some residue of social friction with previous citizens of the area remains. This friction is diminishing, but reconciliation of the two parties is essential for the city's sociopolitical unification. Introduction of a cable television system called ACCS (Academic New Town Community Cable Service) has helped communication between these two groups. In addition, UTV (the television system within Tsukuba University) and JICST (Japan Information Center of Science and Technology) will be completed and unified in the near future. This plan is now under consideration and is more easily implemented here than in other cities because a coaxial cable network was installed in an early stage of the Tsukuba area's construction.

Unification Problem

Politically, Tsukuba still has to confront the unsolved question of unification of all six municipal bodies in this science city site. The recent Tsukuba Science Exposition did not help to boost unification. There is strong opposition among native Ibaraki residents to the merging of these communities into a single city. Tsukuba Science City faces a dilemma. Because there is integration of national research agencies, it would be preferable for the public R&D sector to have R&D venture businesses rather than mass production factories as a symbiotic partner. This, however, would not produce job opportunities for residents. This is an inherent problem inflicted on a high tech–oriented society transplanted to an agricultural society.

Intracity Traffic Problem

Traffic system improvements are urgently needed within Tsukuba. Its public transportation system and private automobiles are inadequate. Bus service is scarce and slow, and less than a half of public residences have private parking lots. Taxi services are too expensive for daily use. Japanese workers are accustomed to bar hopping after business hours, which is not facilitated by the present public transportation services. This custom is felt to lubricate human relationships and to boost professional connections. If these public transportation services are significantly improved, Tsukuba will be close to an ideal city with minimal commuting problems.

Tsukuba/Narita/Tokyo Triangle—Future Traffic Network Development

If further traffic networks are developed in a radial form with Tokyo as the center, then these will accelerate congestion and concentration in Tokyo. Therefore, other approaches must be considered. Traffic network systems should be developed that connect directly with other major developing cities on the outskirts of Tokyo. Commodity market centers and airports must have equal accessibility from any of these cities. In other words, Tokyo and its outlying cities must have equal consideration.

For instance, there is a profound need for a fundamental reorganization of traffic systems on the edges of the triangle of Tokyo/Tsukuba/Narita International Airport. If a rapid transit railway were constructed between Tsukuba and Tokyo, and if urbanization around these areas is well developed, the unreasonable land price gap between Tokyo and Tsukuba would be greatly narrowed. To do this, a second Joban railway is being planned. By the year 2000, this railroad will strengthen the Tokyo/Tsukuba corridor. If Narita and Tsukuba were connected by either express highway or by rapid transit train, then Tsukuba's global importance would be enhanced.

Efforts to Fortify R&D Agencies' Relations and R&D Support Functions

Tesuzo Kawamoto of the Tsukuba Research Consortium notes that Tsukuba lacks sufficient communication among disciplines and that research does not focus on the interests of private enterprises (Kawamoto 1983). How multilateral communications can be fostered, with networking of researchers from various research institutes and universities, is a major issue at Tsukuba. A number of joint scientific study groups have been formed across agency turf borders. These gather periodically. Due to strong budgetary controls in each ministry, no major interagency projects have yet been inaugurated in Tsukuba. A group sharing Kawamoto's concern for integration and cross-pollinization organized the Tsukuba Conference (*tsukuba kaigi*) in March 1987, which focused on interdisciplinary communication and dissemination of Tsukuba's intellectual products, both domestically and internationally.

To assist R&D support functions, the Tsukuba Research Support Center (provisional name, Kenkyu Shien Center) was established in 1987 on a four-hectare (1.6 acres) space close to the city center. This center is funded by the Japan Development Bank, Ibaraki prefecture, and private companies and run as a third sector. The total funding level will be Y6.5 billion ($46.4 million), and the annual expenditure will be Y1 billion ($7.1 million). Further efforts are also being made to establish a new technical high school, where subjects relating to information science will be taught to 250 students in the entering class. The name of the school is Tsukuba Research Academy (*Tsukuba Kenkyu Gakuen*).

IS TSUKUBA SCIENCE CITY
LIVABLE AND PRODUCTIVE?

A survey for the period 1975 through 1980 addressed the question, How do Tsukuba residents feel? In this survey, most Tsukuba residents felt that Tsukuba provided good natural surroundings and spacious housing with functional conveniences. The major problems were regarded to be medical and transportation services and the cultural environment. Among these, medical services and transportation are viewed as the most serious.

Couples with small children felt that Tsukuba Science City provides good school facilities as well as an environment in which children could play freely and safely. University students were not satisfied with the city itself but were satisfied with their university facilities and human relationships in Tsukuba. Others thought in general that the city was comfortable except for frequent traffic accidents. Other complaints included a shortage in administrative services compared with large cities and unnecessary cash expenditures, such as for inconvenient transportation services or costly amusement facilities.

Tsukuba Syndrome

When the national research and educational agencies were relocated, a number of problems were noted among students and researchers, the roots of which seemed to be boredom due to isolation. Researchers' wives had few opportunities existing for window-shopping and other cultural activities and no part-time job opportunities. Improved counseling and cultural programs are seeking to address these stress-related issues.

INFORMATION SOURCES

How Researchers Feel?

Tsukuba is a difficult place in which to obtain necessary scientific information and good research support services. For instance, the National Land Agency (*Kokudo-cho*) conducted a survey of researchers' professional information sources. Over 1,300 were polled on the various sources from which they acquired their essential research information. From within Tsukuba Science City, 8.8 percent received information, from the rest of Ibaraki prefecture 3.4 percent, from Tokyo 23.1 percent, from the rest of the Kantou district 11.9 percent, from Kinki district (Osaka, Kobe, Kyoto areas) 8.3 percent, from the rest of Japan 10.5 percent, and from overseas 14 percent. This shows that most researchers' dependency on Tokyo cannot be ignored.

The 1985 Current Contents Address Directory carries first authors from some 7,000 journals and magazines and 3,500 books published throughout the world. It also gives the city of these authors. The major cities are (1) Moscow (26,000), (2) London (16,000), and (3) Tokyo and Paris (11,000). Tsukuba's author population is less than 2,000; Berkeley (U.S.), 3,500; Stanford (U.S.), 2,900; Cambridge (U.S.), 3,800; and Cambridge (U.K.), 2,900. In this regard, Tokyo seems to play a mother role to Tsukuba.

Five Drawbacks and Three Goals

According to Atsushi Shimokoube (Onda 1986), the five greatest shortcomings of Tsukuba are evidenced in its need for

1. Common orientation among diversified disciplines;
2. R&D flexibility of agencies and potential to stay abreast of innovations;
3. Internationally open systems that welcome participation of foreign researchers;
4. Entrepreneurship that will bring speculative vitality; and
5. Network systems that guarantee efficient transportation and communication and knowledge storage systems like open access libraries and data banks.

Most of the present research institutions seem to move toward institutional stability and rigidity, which does not enhance the possibility for innovation. Shimokoube also outlined three major goals for Tsukuba:

1. Attain satisfactory conditions for commuting and resident workers;
2. Harmonize the natural environment and urban structure; and
3. Create a symbiotic existence of mass systems and individual-oriented systems.

TECHNOPOLIS AND TSUKUBA SCIENCE CITY

Technopolis Development

Recently, the income gap between Japanese city areas and regional areas has begun to narrow. The quality of life in regional cities has greatly improved. This has triggered a U-turn phenomenon of urban residents who originally came from rural areas. The bullet train network, airports with jet runways, express highways, and other high-speed traffic systems have been developed throughout Japan. Therefore, technopoleis aiming at small-and-light (*keihaku tanshou*) product manufacturing seem to be able to take root anywhere. This type of industry is called foot-loose, although heavy (*juuko-choudai*) product manufacturers, which are foot-tight, have settled along coastal areas with ports and convenient energy supplies.

In Japan there are over 200 public research laboratories or test facilities owned by prefectures and major cities. However, these are rather small in scale, and the average number of technical staff is about thirty. Since 1975 MITI has stressed the problem of income discrepancies between urban and regional areas and has installed five local industrial testing centers, each of which has over sixty researchers. The first oil shock in 1973 struck energy consuming industries such as steel and shipbuilding. Under such circumstances, MITI launched the technopolis concept in 1980 (see Table 3–2). Based on this idea, technopolis development legislation was enacted in July 1983. Up to now, about twenty areas around Japan have been designated as technopolis areas. In the near future, twenty-five technopolis areas are to be designated. The aim of the technopolis law is to introduce high tech industries to less developed areas. But R&D organizations tend to be attracted to large cities. In view of this situation, MITI presented a "research core" concept to attract these R&D organizations to technopolis areas by a law enacted in 1986 regarding efficient allocation of R&D staff to technopolis research centers. As of August 1986 twenty-eight research projects were being carried out in already designated technopolis research centers.

Table 3-2.   History of Technopolis Development.

Year	Event
1980	Announcement of MITI's technopolis concept.
1981	Survey report for the Technopolis '90 Construction Plan.
1982	Establishment of technopolis basic plans. Interim report by technopolis committee.
1983	Establishment of technopolis development plans. Law for accelerating regional development based on high technology industrial complex (technopolis law) becomes effective. Promulgation and enactment of technopolis law. Announcement of development guidelines.
1984	Approval of development plans for Kumamoto, Oita, and fifteen other areas.
1985	Approval of development plan of Nagasaki prefecture and three other areas.
1986	Comprehensive economic measures (by conference of related cabinet ministers on economic measures). Confirmation of further promotion of technopolis concept. Economy-stipulating package adopted (at a meeting of cabinet ministers on economic measures). Reconfirmation of further promotion of technopolis plans as the major feature of the rural and regional development promotion policy.

Technopolis Issues

Because local public research and testing agencies are often located in sparsely populated districts and are relatively small, recruiting systems for researchers do not function very well. In addition, public systems are usually not flexible in personnel matters and budget operations. They seem unable to keep pace with the rapid development of modern technologies. Moreover, private R&D facilities tend to gather in the Tokyo area. Can the technopolis law change this tendency? High tech industries in isolated local areas do not attract relevant industries or develop communication with the local society. These are issues that still must be addressed.

On the other hand, major incentives for the attraction of high tech industries are the enthusiasm of local governments, support service systems for high tech R&D, availability of venture capital, guarantees for loans, quality living conditions, and so forth. In this regard, the key issue in Japan is whether

Japanese universities and colleges can support high tech R&D at a satisfactory level. Even if a technopolis proves to be easy to transplant, it is a virtually insurmountable problem to relocate universities or research agencies. In this sense, obvious constraints exist to construction of high tech industry parks for which it is essential to have industry-academic relationships.

In Japan, much remains to be done concerning financial support for these high tech venture businesses. Therefore, to counter disadvantages, there is a great possibility that Japanese high tech industries may find a home somewhere outside of Japan in the future and that foreign high tech R&D ventures may find a way to settle in Japan.

Cooperation from Universities

Transferring urban universities and research agencies to regional areas is a great bottleneck in the construction of technopoleis. In this sense, Tsukuba is the sole exception. Although it would be too much to expect actual investment by universities in this country, private enterprise wonders to what extent universities could be cooperative in joint R&D or how universities might provide consulting services. For example, could one expect university lectures through CATV for workers at their business places for credit, as is done at Stanford University?

In this regard, the International Science Foundation (Kokusai Kagaku Zaidan) was established by the Ministry of Education at the University of Tsukuba for the betterment of industry/academic relationships. Still, other problems remain, like how venture businesses or private universities can be attracted to Tsukuba.

Venture Capital

As for venture capital problems in Japan, there are debt guarantee companies and investment companies. But contract conditions are much more severe than in the United States. Therefore, those companies are not effective in providing necessary capital to emerging small venture businesses that need investment. This is a genuine concern.

Tsukuba as Technopolis

In the Tsukuba area, high tech industries accounted for 20 percent of all industries at the beginning of 1984, an increase from 5 percent in the period

1975 through 1980. The high tech percentages have continued to grow since 1984. The major incentives for attracting these industries have been easy labor recruitment and policies by local governments for quality of life, comfortable climate, cooperative research universities, affluent national security R&D with favorable procurement budgets, venture capital, and other important factors. Modern types of production have shifted from manufacturing goods to value-added manufacturing. In this sense, the hope is that if knowledgable people are brought together in a specific area, the result would be to enhance intellectual productivity. But this does not necessarily mean that Tsukuba attracts venture industries because land prices and labor costs are proving to be a disadvantage for labor-intensive industries.

Who Wants What?

Expectations for Tsukuba will obviously vary according to the perceptions of different groups—researchers in public R&D agencies, city planners, new residents. From a city planning point of view, Ibaraki prefecture certainly should have more industries that can employ more people and can produce more sales, as opposed to R&D agencies. Other prefectures that have a technopolis or are going to construct a technopolis within their territory would not like to see the concentration of industries in the Tsukuba area. They would rather have R&D agencies. There are, in other words, differences between Tsukuba's development and that of other technopoleis.

CONCLUSION

It is still too early to judge the success or to evaluate the state of Tsukuba Science City and Japanese technopolis strategy. The technopolis concept sprang initially from a concern about the population distribution problem and urban/rural problems rather than from a desire for pure science and technology promotion policies. These two drivers have different requirements and implications. The key to the success of the technopolis is how effectively city planners and science and technology policymakers can work together and lead the cities to more productive states, which might prove to be totally different forms from what has ever existed before.

NOTE

I am very grateful to Professor Hideo Sato of Tsukuba University, who extended me the opportunity to write this chapter.

I would like to thank Professor Shuntaro Shishido of Tsukuba University and Director Tetsuzo Kawamoto of the Tsukuba Research Consortium for providing me with useful materials and data to construct this paper.

Above all, I am greatly indebted to Makoto Kobayashi, director general of the planning department of MITI's Small and Medium Enterprise Agency, who arranged a field study tour to major technopolis areas. I wish to thank other MITI officials who provided me with useful information on the technopolis and on MITI's industrial R&D strategy. These include Masahiro Omura, Ryuji Yoshino, and Takashi Inoue for discussions on technopolis issues.

I am also indebted to Reverend Harry Burton-Lewis, who has lived in Tsukuba over fifteen years and was kind enough to give thoughtful suggestions and to glance over my English.

BIBLIOGRAPHY

Burton-Lewis, Harry. 1985. "Science City in Transition," *Look Japan* 30 (348).

Current Contents Address Directory of Science and Technology. 1985. Philadelphia.

Future Images of the Brain City. 1984. Symposium reports from the Seventh Meeting of the Fifth Subcommittee of the Japan Administration Planning Society, Tsukuba.

Hiramatsu, Morihiko. 1986. "A Challenge to Technopolis" (*Tekunoporisu eno chousen*), *Nihon Keizai Shimbum*.

Ibaraki Techno-Linkage Concept. 1985. Planning Division of Ibaraki Prefecture, March.

Imai, Kenichi. 1983. "Japanese Industrial World" (*Nihon no Sangyo Shakai*). Chikuma Press.

Industrial Allocation Vision in the 21st Century (*Niju-isseiki no Sangyo Ricchi Bizyon*). 1985. Tokyo: Industrial Location Guidance Division, MITI.

Kawamoto, Tetsuzo. 1983. *Tsukuba as a Research Center.* Environmental Study of Tsukuba Science City, Environmental Study Group in Tsukuba University.

Kawamoto, Tetsuzo, and Tokio Wakabayashi. 1984. "Research Oriented Private Enterprises in Tsukuba Science City." Environmental Study Group, Tsukuba University.

Mark, Hans, and Arnold Levine. 1984. "The Management of Research Institutions." NASA SP-481. Washington, D.C.: NASA.

Ogata, Kenjiro. 1986. "Policy for Technopolis Construction." A paper presented to the International Association of Science Parks' Meeting, Kumamoto.

Onda, Masahiko. 1986. *Tsukuba in the Future (Ashita no tsukuba)*. Selected presentations in Tsukuba Science City compiled by Tetuzou Kawamoto. Step Press.

Tatsuno, Sheridan. 1986. *The Technopolis Strategy*. Englewood Cliffs, N.J.: Prentice-Hall.

Yoshino, Ryuji. 1986. "Japan's Technology Development Based Regional Development." A paper prepared for NBST Expert Workshop Research and Technology Development for Less Favoured Region. Tokyo: MITI.

Chapter 4

A NEW RESEARCH INSTITUTE IN CHINA

Yu Jingyuan, Song Yuhe, and Zhou Zheng

This chapter describes recent Chinese reform policies concerning the management of scientific and technological institutes and exemplifies these changes through a description of the Beijing Institute of Information Control (BIIC). Developing trends of Chinese research institutes and policy problems resulting from these reforms are also discussed.

REFORMING CHINA'S SCIENCE AND TECHNOLOGY SYSTEM

China's current system for the management of science and technology, which came into being under specific political conditions, has demonstrated its strength in concentrating resources and effort to solve major scientific and technological problems. But this system is increasingly falling short of meeting the requirements of the four modernizations: industry, agriculture, defense, and science and technology. The major problem lies in the tendency to isolate science and technology from production.

After the Third Plenary Session of the Eleventh Chinese Party Central Committee at the end of 1978, the Chinese government implemented an open policy and launched its economic reform plan. As China's rural economic reforms met with success, policy members turned their attention to urban economic reform and political system reform. A further goal is now to encourage the technological revolution by dovetailing science and technology with production, thus achieving coordinated development of science and technology, and the economy and society.

Scientific and technological work should serve economic development. However, in China science and technology have long been separated from the nation's economic progress. Research institutes have been answerable only to their superiors and lack any ties with factories or business enterprises. As a result, a great number of technical problems arising from production are not solved, and a majority of research findings are not widely utilized in production. This situation results in drawbacks in both the economic structure and the development of science and technology. Drawbacks in science and technology can be summarized in the following four areas:

First, in the past most of China's research institutes have been completely funded by the state. Researchers did whatever they were told by higher authorities who paid little attention to the research needs of the real world. Although research recipients received the research results free, these findings were often not applicable to the problems at hand. As a consequence of such a policy scientific and technological research has had a minimal effect on economic development.

Second, under the old policy the five segments of China's science and technology system—the Chinese Academy of Sciences, the Education Commission, research units under local governments, and the military and civilian ministries under the State Council—were severely cut off from each other. As a consequence researchers and technicians have been severely restricted in their work, the reasonable flow of talent has been stymied, and the key areas of research, design, education, and production have been disjointed.

Third, government control and management of science and technology has been too rigid and all-embracing. Research institutes have lacked the capability of self-development and the initiative to serve economic construction.

Fourth, economic enterprises have lacked the capability of using science and technology results. There are no technology commercialization markets available in China. Intellectual work has not been properly valued or respected.

To redress these situations, in March 1985 the Party Central Committee of China issued the decision to reform the Scientific and Technological Management system. To this end, over the past year the Chinese government has been attempting to implement the following series of reform policies:

1. To replace the traditional state funding system for research institutes with localized fund management systems. State research funds will be gradually reduced, but not canceled, until the institute can support itself.

2. To speed the commercialization of research results by sponsoring technical information fairs, opening up channels for the smooth flow of research findings into production fields, and changing the practice of free transfer of technology achievements.

3. To facilitate the ability of enterprises to incorporate advanced technology and develop new products. Different technological development research activities will be encouraged to cooperate with enterprises and institutions to establish a variety of interassociations with the goal of gradually merging these associations into one economic entity.

4. To reform the management of technical personnel so that scientists and technicians will be permitted to take on extra jobs and to work for other institutes reasonably freely. The idea is to promote the integration of scientific research with production and to give full scope to the talents of researchers and technicians.

These policies are being gradually implemented by research institutes. For instance, with changes in the funding system, the operational funds formerly granted by the state will gradually be reduced, and this will place heavy economic pressures on most institutes. Because this one reform involves over 6 million scientists and technicians and more than 9,000 institutes across the country, prudent steps are required to avoid disruption and losses. At the same time, people must blaze new trails because there is no Chinese-based model to follow. Beijing Institute of Information and Control (BIIC), for example, has instituted some reforms and its actions have been highly commended by the government.

AN EXAMPLE OF AN INSTITUTE UNDERGOING REFORMS

BIIC was established in 1982. Professor Song Jian, state councilor and the current chairman of the State Science and Technology Commission of China, was the first director of this institute, which has been quite successful in carrying out the reforms in economic, scientific, and technological management systems. By the end of 1986 BIIC employed 376 staff members (including 185 scientists and engineers), had established technological links with nine foreign research institutions and 150 domestic enterprises, and had signed more than 250 contracts with various Chinese departments and local governments. The institute earned an income of nearly 7 million yuan in 1985 and 14 million yuan last year (see Table 4–1). The per staff income reached 40,000 yuan in 1986, twenty times the nation's average.

Table 4–1. Sources of Funding by Year and Staff Levels for Beijing's Institute of Information and Control.

Year	Government Funds	Income	Staff Level
1982	2,287,300 yuan[a]	12,652 yuan	99
1983	4,221,000 yuan	55,407 yuan	209
1984	1,111,000 yuan	1,146,428 yuan	267
1985	1,778,500 yuan	6,883,566 yuan	328
1986	1,037,000 yuan	14,039,050 yuan	376

Source: Beijing Institute of Information and Control, 1987.

[a]1 yuan equals $1.00.

BIIC includes two major organizations: the Computation Center and the Research Center of Systems Engineering. The Computation Center consists of three functional departments and five laboratories as follows:

1. The Computer System Management and Operation Department:

   Operation and management for mainframe and minicomputers
   Environmental engineering for computers
   Computer network implementation, maintenance, and management
   Computer security
   Computer system configuration and disposition
   Computer reliability, maintenance, and availability
   System software development
   Spare parts supply, system maintenance, and technical training for Burroughs computer users and other services

2. The Information Service Department:

   Programming
   Algorithm studies
   Consulting
   Technical training

3. The Laboratory of Computer Networks:

   Techniques of local area networks
   Wide area network systems

Network interconnection techniques
Networking techniques for computers with different protocol transfers
Network technique for remote and real-time information monitoring

4. The Laboratory of Data Base and Information Management:

Development of data base techniques
Analysis, design, and implementation of information management systems
Transaction-oriented high-level (fourth-generation) languages

5. The Laboratory of Application of Computer-Aided Engineering:

Computer-aided structure design
Computer-aided control system design
Engineering data base
Intelligent work-station

6. The Laboratory of Image Processing and Pattern Recognition:

Image processing and pattern recognition techniques
Three-dimensional image generation, reconstruction, and display
Character recognition
Application development in remote sensing, industry, and medicine
Real-time transmission and processing systems for image and signal
Artificial intelligence

7. The Laboratory of Microcomputer Applications:

Application of microcomputers to design automation
Application of microcomputers to information management automation
Local network of microcomputers and office automation
Application of microcomputers to process control
Microcomputer simulation

8. The Department of Information and Documents:

Research and analysis of computer application information
Collecting and offering of home and foreign computer application
    information
Publishing specially collected works and translations

Filing technical reports
Care of books and reference materials
All kinds of document services
Academic exchanges

The Research Center of Systems Engineering consists of five laboratories, which are conducting laboratory research in the following areas:

1. The Laboratory of Systems Science:

Theory, methodology, and application of systems science
Intelligent robots
Control theory and differentiated dynamics

2. The Laboratory of Economic and Environment Systems:

Decision analysis and quantitative analysis of macroscopic and microscopic economic systems
Environmental protection

3. The Laboratory of Science and Technology Systems:

Economic structure
Technology marketing
Science and technology development policies

4. The Laboratory of Social Systems:

Population modeling
Population prediction
Simulation modeling of regional population planning

5. The Laboratory of Application Software for Economic and Social Systems:

Data bases, model bases, and method bases
Satellite data communication
Information access service

Funding System Reforms

Traditionally Chinese research institutes are assigned tasks by the state and receive funds to do the work from the government. Because BIIC was a new institute that lacked advanced equipment and famous researchers, it received neither tasks nor substantial research funds from the government. Also, what little government support existed has been gradually reduced. Consequently, the institute and its managers have been under heavy pressure to find paying customers in need of scientific research.

Initially research at the institute concentrated on computer science and computer engineering including information processing, scientific computing, computer-aided engineering applications, system engineering, and other research work in relatant fields. BIIC also provided information and comprehensive services on computing for nationwide factories, government organizations, research institutes, universities, and so on. The expansion of this research directly to the users was an immediate success. Over the past five years the institute has taken on more than 220 research contracts. Although twenty-eight were awarded by the government, the rest were obtained by the institute from enterprises. In this way, the institute has supported itself.

In 1984 BIIC initiated "soft science" research (or policy research), which included economic analysis, population development prediction, scientific and technological policy analysis, and environmental protection research for the national government and ministries and the local governments of provinces. Over the past three years, the institute has completed twenty-one "soft science" large research projects for the government independently or jointly with domestic and foreign research institutions. Some of the these projects are National Dynamical Macroeconomic Control Modeling supported by the State Planning Commission, which was used by the State Council in drafting the seventh Five-Year Plan (1985–90); a Comprehensive Equilibrium Model of Financial Subsidies, Prices, and Wages for the State Economic System Reform Committee; Analysis and Optimization Model of the Monetary System for the Bank of China; Price-Tax-Finance System Model for the Finance Ministry; data bases, model bases, and method bases for the State Economic Information Center; specifications for the 1986 Beijing-Vienna Satellite Data Communication Line, the first worldwide information access system in China; R&D Cost Analysis and Planning Models for the Defense Research Commission; and Population Control Research in cooperation with the East/West Research Center in Hawaii. These research projects have provided scientific support for Chinese decision-makers and have provided BIIC with enough surplus income to pay employee bonuses and to fund additional scientific research.

Responsibility Systems Take Hold

Researchers and technicians in most Chinese institutes do not care much about how they spend funds. They do not give much thought to efficiency when purchasing equipment and material. Their only concern is the research result. Waste and overstocking are rampant, and overstaffing is common. But the institutes do not worry about these problems because the state pays the money. Furthermore, once the research result is accepted by the state, the institute has no more responsibility for the project. The state assigns research tasks, allocates funds, makes up for losses, and, of course, takes all the profit if there is any.

But this is not the case with BIIC. To bring vigor and vitality to the institute, a "director responsibility system" has been fully implemented. The director takes full responsibility for management and for profit or loss and cannot afford to overstock or overstaff. However, because it is impossible for the head of the institute to attend to every detail, the institute has implemented a system in which each staff member takes responsibility for his own work.

BIIC is divided into several research groups, and the members of these groups can be from different laboratories. Each group is responsible for its own research projects and for using its income to buy needed research equipment and material. Part of the money that they save can be used as profit and distributed as bonuses to group members. In short, each research group has the power to manage its own business.

These groups plan carefully before purchasing instruments or material. The more the group spends, the less the profit and the lower the bonuses. Because a person's performance is directly linked to his economic reward and the available surplus funds, initiative is stimulated and resonsibility for research results is strengthened. Each individual in each group has the responsibility to make sure the research results are delivered quickly and that they are a quality product. Otherwise, the group's income would be reduced.

Academic Exchange and Personnel Management Reform

BIIC has been trying to expand international and national academic exchanges to increase the number of foreigners studying in China and the number of Chinese studying abroad and to invite foreign specialists and other scholars to work at BIIC. At the Third National Conference on China's Association for Science and Technology in 1985, Professor Song Jian, the first director of BIIC, said that there was no model for the Chinese to follow in developing the high technology necessary to keep up with the world's scientific and technical developments. Accordingly he recommended that China should strengthen international exchange programs. Since the beginning of BIIC, its basic operating policy has emphasized openness to the outside world and cooperating with others. In recent years the institute has established various forms of international and nationwide academic cooperative relations, hosted fourteen national academic conferences and two international academic symposia, and joined three world and five national scientific organizations. Such a policy of openness has greatly contributed to the success that has been achieved by BIIC in narrowing the gap between Chinese science and that of the more advanced nations.

Emphasizing on-the-job training for staff has been another basic policy of BIIC. Over the last few years more than twenty people have been sent abroad to study for advanced degrees or as visiting scholars. More than one-third of BIIC's staff members have been accepted for training in English and other majors. Also, since 1984 BIIC has had more than fifty employees registered as graduate students.

Furthermore, BIIC has conscientiously helped its staff members improve their living and working conditions. In recent years, two blocks of flats, an apartment house, and an office building have been constructed. Through such reforms, BIIC has created a better environment for scientists and engineers so that they may develop to their full ability.

In summary over the past five years, BIIC has carried out comprehensive and systematic reforms in scientific and technological management. These actions have received the earnest support of many government leaders, including Premier Zhao Zhiyang, State Councilor Son Jian, Vice-chairman of China Science and Technology Association Qian Xueshen, and the famous economist Ma Bin. Experience has proven that reform policies such as the contract system, the responsibility system, and an open policy can be successful in China. Such direction is correct and is being adopted by others. For example, according to the State Science and Technology Commission, in 1985, 40 percent of China's technical development and research institutes have installed the contract system. On the other hand, these changes are taking time. For example, even innovative BIIC is still involved in replacing the old system with the new.

THE FUTURE DIRECTION OF CHINA'S RESEARCH INSTITUTES

After the publication of the Central Committee's Decision, research institutes across the country have begun reorganizing and adjusting their operations to be in line with the decision. The most important question for the directors of all the institutes has become how to get sufficient research funds given the decrease in funding from the past channels: major state scientific research projects, contracts with various departments and local governments, and funds from science foundations. The new operating conditions, coupled with Chinese scientists' generally strong desire to contribute to the revival of the national economy, is increasing the scientific and technological work that contributes to the country's economic development.

Recently, the State Council issued "Regulations to Promote Reform in the Scientific and Technological Management System." According to these regulations, every ministry under the State Council should devolve themselves of administrating scientific research institutes. Rather, they should exercise indirect management of scientific research units through policy guidance and coordination. Most of the scientific research units that focus on research and development of new processes and products should gradually unite with manufacturing firms. Research funds should eventually come from the institutes' earnings from the enterprises that purchase their research results. Research institutes should also adopt other ways to serve the national economy and their funds for research and development should also depend on income from these services.

It is believed that such policies will promote the integration of industrial enterprises and scientific research institutes, accelerate commercialization of technological achievements, and expand the market for technological transfer. As a result, the majority of Chinese research institutes will be combined with large and medium-size enterprises. And these enterprises will be stronger in production and management and capable of developing new products. Although this is the main direction for China's research institutes, the government will still grant a fixed amount of funds to those institutes engaged in public welfare activities and technical services as well as multidiscipline research projects advanced by scientists and technicians. Basic research will be concentrated in the universities and some research institutes under the Chinese Academy of Sciences supported by the Science Foundations.

In order to speed the reform of scientific and technological management systems the following problems still need to be solved. First, the most popular products at the technical trade fairs have been the simple, time-saving technologies that are most certain to yield quick economic returns in the process of production. Much less favored are the more sophisticated and time-consuming technologies that might involve production risks even though these technologies have great protential to increase production and reduce cost. Such practices indicate that the buyer's market at the technological fairs have yet to be fully developed. This will require upgrading the technology of large- and medium-size enterprises through urban economic reforms, as well as the growth of demand by all levels of enterprises for research findings, which will enhance production.

Second, according to regulations, by 1990 the state plans to decentralize its administration of nearly 5,000 research institutes, which will be combined with the country's 7,000 large- and medium-size enterprises. A serious problem inhibiting such a merger is that the institutes are operated differently than the enterprises. Research is needed on how to merge the research institutes with the production and management requirements of the enterprises.

Third, in China, scientific and technological researchers are still not fully appreciated. One important indication of this is the low investment by the state in basic and multidisciplinary scientific research. In the Chinese Academy of Sciences for example, there are 22,000 senior and middle-level researchers, accounting for 69 percent of the academy's total researchers. All of these researchers, who have an average age of 43, have a strong desire to work hard and to engage in independent research; however, because investment in scientific research averages only about 9,000 yuan per researcher, such contributions have been slowed and researchers' initiatives have been stifled. Because it is not currently practical to get more investment from the state, the only alternative, outside of support from business enterprises, is to concentrate the available funds on just a few of the most worthwhile research topics. One result will be that some research personnel will have to be transferred to other jobs because they cannot get the research funds to do their own work. Consequently, a key problem in the course of the current reforms is how to increase the prospects for employment for these personnel.

In an effort to avoid overstaffing and waste of talent the state has suggested that technical personnel might be mobilized to work in medium-size and small cities, rural areas, remote border areas, and communities of ethnic minorities. The implementation of such a proposal, however, would involve many difficulties because such areas generally lack the favorable social and scientific support conditions needed for research. It will take a major policy effort to create conditions in such areas, which will be attractive to the research workers.

Finally, as the state stipulates, while promoting technological development, efforts should be made to increase research in the applied sciences as well as bring about sustained, steady development of research in the basic sciences. However, in order to realize these goals, it is necessary to establish a more complete and supportive funding system.

BIBLIOGRAPHY

Decisions of the Reform of Scientific and Technological Management System. 1985. Beijing: Party Central Committee of China, March 13.

Regulations to Promote Reform in the Scientific and Technological Management System. 1987. Beijing: State Council of China, February.

Song, Jian, and Jingyuan Yu. 1985. *Population Control Theory*. Beijing: Science Press.

Yu, Jingyuan, and Yuhe Song. 1985. *Journal of Systems Engineering* (Human) 12.

Yu, J., Y. Song, and S. Yang. 1986. *Quantitative & Technical Economics* (Beijing) 1.

Yu, J., and others. 1986. Proceedings of IFAC Workshop on MDG, Beijing, August.

Yu, J., Y. Song, and Z. Zhou. 1986. Proceedings of Twenty-fifth IEEE Conference on DC, Athens, December.

Chapter 5

THE CAMBRIDGE PHENOMENON
Universities, Research, and Local Economic Development in Great Britain

N.S. Segal

The Cambridge phenomenon is the nearest (although still small-scale) European equivalent of Silicon Valley and Route 128, with small local R&D-based firms spearheading economic development. The phenomenon in its totality is unplanned and its origins multiple. The university's presence, research excellence, and hands-off policies have been crucial to emergence of the phenomenon. Public agencies have played no direct role, but sustaining its long-term growth will require conscious and systematic effort by many different interest groups. Although there are other models of technopolis and different routes to them, the Cambridge phenomenon stands as especially significant in the British context.

The Cambridge phenomenon is the name given to the burgeoning growth of high technology industry, led by locally formed small firms, in and around the university and market town of Cambridge, England. Awareness of the phenomenon has coincided with a period of intense interest in university/industry research relations, in the (separate) roles of universities and of small firms in local economic development and in the importance of science and technology in the overall economy. Partly because of this coincidence the Cambridge phenomenon has attracted national and international attention.

But it is important to recognize that the phenomenon, much as it bears on these and related topics, is a complex development process whose origins and characteristics are unique to the specific circumstances of Cambridge over a long period of time. Other places and institutions have their own unique circumstances, opportunities, and constraints. Neither Cambridge nor any-where else stands as a universally applicable model. As the phenomenon matures, it is becoming increasingly evident that factors in addition to those that caused and shaped its emergence will bear on its ability to sustain its growth into the long-term. Again, these circumstances are peculiar to Cambridge; different factors may apply elsewhere.

EMERGENCE OF THE PHENOMENON

The Cambridge phenomenon represents a growth of high technology industry on a broad sectoral base in which the leading role is being played by start-up and growth of small, locally formed and independent businesses. These firms are engaged mostly in research/design/development or in high-value, low-volume production; what bulk production there is is subcontracted elsewhere.

The phenomenon's origins are long-standing. The first two high technology companies (which still exist, although in different forms) go back to the nineteenth century. And some of the main factors that have given rise to it can be traced back many decades, at least to the midnineteenth century. But significant growth is recent. Around one-half of the over 400 high technology firms in existence in early 1986 had been set up after 1980; in 1974 there were perhaps 100 firms in total and thirty in 1960. Employment growth has been of the order of 8 percent a year since the late 1970s, virtually all of it coming from firms set up in that period. The overall impact on the labor market has been even greater because of a high employment multiplier estimated at 1.0.

The phenomenon is not (yet) large in scale. The population of Cambridge itself is only 100,000, and of the total labor catchment area some 250,000. The high technology firms, which are concentrated in and close to Cambridge, accounted for 16,500 jobs in early 1986. This was equivalent to 12 percent of total employment.

The small firms have typically been set up by more than one individual "spinning out" of existing local firms (the largest source) and of other local organizations, including the university. The clustering of small high technology firms has in turn attracted new operations of large companies to the area, both those engaged in advanced technology and in financial and business services. There has also been a boom in the growth of the small-scale financial and business services sector, due to and in turn reinforcing growth of small high technology companies.

Small-scale and young as it is in absolute terms, the phenomenon nevertheless is thus far the nearest European equivalent to Silicon Valley and Route 128.

It is a homegrown and spontaneous economic development process, unplanned in its totality, the outcome of a large number of essentially independent but mutually reinforcing events and decisions over many years.

CAUSE OF THE PHENOMENON: UNIVERSITY INFLUENCE

Any such development process inevitably has multiple causes, some long-run and preconditioning and others short-term and precipitating. Two that stand out as profoundly important are the role of the university and the urban and industrial context in which the development has taken place.

The university's influence, while mostly indirect, has been central to the origins and early stages of the phenomenon. Indeed, the phenomenon would not have arisen had there not been a university and more particularly a university of the distinction and style of Cambridge. There have been three components to the university's role.

The first derives simply from the presence of the university and its historically pervasive influence on most aspects of life in Cambridge. It is because of the university that sizable numbers of research scientists live in the area and that there is a continuing vitality about the local cultural, intellectual, and social science. It has created an environment in which highly qualified individuals and their families—be they scientists, engineers, other professionals, or managers—like to live and that offers access to, for instance, excellent educational amenities.

In addition the university has set a tone and style of quality, individualism, and confidence; to some this can be arrogant, but even so these qualities are essential to entrepreneurship. Because of the collegiate structure, the university has provided a unique environment for social and interdisciplinary contact within and between the entire academic and research communities and increasingly now also the high technology and business communities. The concept of networking helps explain how Cambridge has functioned as a university and market town in the past and now also as a high technology business center.

The second component of the university's influence is a product of its international excellence in research, especially though not exclusively in fields in science, medicine, and engineering, which have generated new business opportunities in the past decade in particular. Being a center of internationally recognized excellence, able to attract some of the top brains in the field, has been a signal feature of the Cambridge scene. The number of Cambridge Nobel laureates is out of all proportion to the size of the institution.

The sectoral composition of the high technology firms in the Cambridge area can be traced back reasonably clearly—directly so in some cases—to research conducted in the university and associated institutes. Over a long time, some of the influences have become somewhat obscured, with accumulation of skills derived from the original expertise finding application in new technologies, markets, and products. Separate evolution over many decades of the scientific instruments industry and of computing, and their various conjunctions in electron scanning microscopy, image processing, microcomputer design, and other fields, illustrate the cumulative causation and continuity of development characteristic of growth in a small local economy.

The impact of high-quality academic research in Cambridge has been demonstrated through a number of mechanisms. One, for instance, is where a group of international standing has attracted a flow of bright young researchers on short-term contracts, a few of whom have subsequently wanted to stay in the area and thus have set up businesses themselves. Another is where the leading researchers have acted as consultants to larger companies, some of which have subsequently set up specialist operations nearby. Or again, technicians and software programmers and others have learned specialist skills in support of the primary research group and have thereby added to the pool of expertise in the local labor market. Only very occasionally, it may be noted, has commercialization of research proceeded in a linear fashion originating in the particular research itself.

The third ingredient in the university's influence is a consequence of the loosely specified terms of employment of faculty and the university's liberal attitude toward the ownership and exploitation of intellectual property. The university regards intellectual property as belonging to the individual (who can, of course, give up his rights) and is "laid back" about how its staff deploy their time and effort and to whom the benefits of outside activities accrue. There are two main provisos. First, staff colleagues and students alike must feel that in each case the individual faculty member concerned is still pulling his weight in respect to his teaching, administrative, and research responsibilities. Second, the risks of commercialization fall on the individual not the university, and the latter expects only to be paid for whatever resources if any are directly used in the commercialization process.

This hands-off attitude must not be allowed to conceal the university's concern that wherever feasible academics should engage in outside work and forge links with industry. Such links are seen as both reinforcing and sharpening mainstream teaching and research. Again, there is no compulsion in this.

In general therefore, the university operates a benign, supportive, and noninterventionist policy. To some extent this approach makes a virtue of necessity in that the university is, in a strict sense, ungovernable. It has a collegiate structure with more than thirty independent and financially autonomous colleges, as well as many powerful departments with strong individuals at their head. Most faculty members hold both college and university posts. A central authority could create as many rules as it liked, but it could not monitor them and it could not enforce their implementation. In these circumstances a highly decentralized policy that basically thrusts responsibility on to the individual is both sensible and necessary.

In sum, the university has created a culture of excellence and openness: excellence in its research, openness in its dealings with the outside world and in the dealings permitted and indeed encouraged on the part of its individual staff members. This culture has had the unplanned consequence that, where individuals wanted to commercialize their know-how through company formation, they could do so without having to pretend to the university that they were not devoting their energies and time somewhere else. They could do so overtly, and in a few distinctive cases even initially inside the university laboratories as a temporary home. The university therefore has in effect helped "seed" the process of new firm formation by allowing it to be respectable for academics to exploit their knowledge and make money out of doing so, and at the same time still retaining their academic position if they so wished.

It is the indirect rather than the direct role of the university that has been of special significance. This is clarified and emphasized by a "family tree" of the high technology companies, which shows inter alia how already established firms, the university, and other local organizations have singly and jointly spawned new local businesses.

This genealogical analysis shows that up to the mid-1980s only some fifty-five of the high technology businesses in the area had been set up by individuals coming straight from the university (or still remaining in it). However, the university (chiefly the physics, computer, and engineering departments) has indirectly been the ultimate origin of virtually all of the other companies. This is because first generation spin-outs from the university have themselves spawned new companies, and so on. Even where the parent companies (or other companies that have not yet become parents of spin-outs) have not come from the university, the latter has constituted a central reason for the organization concerned to be located in Cambridge in the first place.

CAUSES OF THE PHENOMENON: URBAN AND INDUSTRIAL CONTEXT

The second major influence on emergence of the Cambridge phenomenon revolves around Cambridge's short and limited industrial history and the town's relative remoteness and small size in a growing region. The essential elements of this are several.

First, industrial market opportunities—generated by demand originally mostly from the university, more recently by already established larger local firms—have been more easily identifiable locally. In the absence of local firms to whom know-how could readily be licensed, these markets have been open to new firm penetration and have been an important stimulus to start-up.

Second, the fact that there has never been heavy industry, or industries in which large plants and large unionized labor forces have been prominent, has helped create a labor market and a general attitude in which flexibility and individualism have never been suppressed. A history of low wages— due to the long dominance of the agricultural and low-level services sectors (the latter partly a result of employment patterns in the university and colleges), reinforced by the early industrial employers—and a generally low penetration of trade unionism, have contributed to efficient functioning of the labor market.

Third, there is no question of the university's becoming engulfed and inconspicuous within a large metropolis. Similarly, it has been much easier than it would in a large city for a "critical mass" of high technology firms to be reached. There are numerous interlocking networks of talented, influential, and accessible individuals, that make for informal, congenial, and efficient business dealings. All these circumstances have allowed the firms to be noticed—in effect the phenomenon to be recognized—with the attendant benefits of practical support and generation of confidence among themselves and among the financial, business, and academic communities in the firms. In London, say, these benefits are harder to realize because high technology firms and higher educational institutions themselves tend to be "lost" or at least their collective impact minimized.

Fourth, on a wider spatial scale the East Anglian region in which Cambridge is located has for several decades been the fastest growing region in Great Britain in population and employment. The starting base was low—a dispersed, mainly rural population with few large towns and no sizable cities—and even now the population of the whole region is not much more than 15 million. Growth has proceeded essentially through the in-migration of people, an above-average proportion of whom have subsequently set up their own firms, and the immigration (subsidiary operations or relocations) of existing firms based elsewhere.

The size composition of firms—in-moving, existing, and new—has been and remains skewed to the small-scale, in part because of the limited industrial infrastructure and the small supplies of labor. To a certain extent, the phenomenon represents what has been going on across the region as a whole; through the rate of new firm formation, the high incidence of high technology firms among them and the geographic concentration of development are distinctive features of Cambridge.

Fifth, closely connected with the growth of the region has been substantial investment in its communications infrastructure. In the past decade, this has been mainly in the motorway and other major road systems, of which Cambridge has been a signal beneficiary. Development now underway of London's third airport at Stansted, twenty-five miles south of Cambridge, and electrification of the London/Cambridge railway line, will together greatly augment these infrastructure developments.

It is important to note that neither central nor local government has played a direct role in emergence of the phenomenon. The center's various support schemes for small firms and for technological innnovation have at best had a modest impact. (But of course, as an indirect influence, central government expenditure on research and teaching have heavily underpinned Cambridge's international excellence in these respects.) And local government, up to now involved chiefly on the physical planning and development side, has essentially been responsive to the pressures of growth and indeed has often sought to restrain rather than promote or shape growth.

Similarly, the Cambridge science park, successful and internationally recognized as it is, is not itself the phenomenon or even a principal cause of it. The science park, which houses some 15 percent of the high technology firms, has certainly played a number of critical roles in the emergence of the phenomenon since the park was first mooted in the late 1960s and opened in 1973. But it must be seen as essentially the flagship of the phenomenon, the visible symbol to both the outside and the inside (university) worlds that something special is happening by way of technology-based development.

SUSTAINING GROWTH OF THE PHENOMENON

The fact of emergence of the phenomenon is exciting, all the more so because of its unexpectedness. That a leading center of high technology industry should emerge in a medieval market town, which accommodates one of the world's famous universities, set in a somewhat bleak even if prosperous rural environment, in a location that has only recently started being well served by strategic communications links, and that historically has been far from the main centers of industrial development, is not something that could easily have been foreseen.

Having emerged and indeed grown impressively in recent years, the crucial question about the phenomenon now is, can its momentum be sustained? There is of course no simple yes-or-no answer, but it is useful to consider what issues bear on the matter. Interestingly, they revolve less around the entrepreneurial energy and business management skills in the local economy, especially among small firms that are growing rapidly. Rather, they relate more to the two topics examined in the previous section: the university and the economic and physical context of Cambridge.

To take first the question of the university. In relative terms, the university has diminished in importance in Cambridge as the phenomenon has grown. Nevertheless, it will remain of fundamental consequence to the companies in the phenomenon and indeed to the area as a whole that the university maintain its excellence within an international, and not just national, frame of reference. One main factor is now putting this excellence at risk. This derives from the financial stringencies imposed by the government on the university and research sectors nationally, from which not even Cambridge has escaped. Morale among university teachers and researchers is low, and there is a rising loss of outstanding individuals to other jobs, most conspicuously in the United States. (In Cambridge there is also an equivalent loss of technicians to the local high technology firms.) Correspondingly, it is difficult to recruit outstanding people from outside, a difficulty compounded by the fact that relative to university salaries Cambridge is now a high-cost place to live.

The university has, of course, limited power to redress these adverse national developments. But the potential for it to mount special initiatives to protect and even enhance its recognized strengths does exist. A few other academic institutions (less well resourced and prestigious than Cambridge) have shown to great effect just what can be done in these respects. The problem faced by Cambridge University in this regard arises from the same circumstances that helped give birth to the phenomenon. The university's devolved and fragmented structure, and the limited powers of the central authorities, make it extremely difficult to take concerted actions. A corollary of the university's being ungovernable is the difficulty of conceiving and implementing strategic initiatives. It is notable in this respect that the university has not so far sought to capitalize in an ambitious way on the fact of the phenomenon and on the potential for new teaching and research initiatives that this has created.

The second main factor effecting the future of the phenomenon is its economic and physical context. It was seen earlier that "accidents" of history and geography—limited previous industrialization, small urban size, dominant role in the subregion, and so on—helped create the conditions in which the phenomenon emerged. These factors are, however, also constraints on growth of the phenomenon. For instance, the small labor market exacerbates the skill shortages that seem inevitable in any high technology development. Similarly, the relative absence of production and management skills contributes to a limited capability to capture all the benefits locally of successful growth of small R&D firms. Route 128 illustrates the benefits that can follow from there being a large and industrially skilled labor force already in the locality.

The small size of Cambridge, especially because of its poorly developed physical infrastructure that was perfectly adequate for the city's traditional purposes until very recently, imposes a constraint of another kind. Severe pressures have developed on the housing and labor markets and on traffic as well as shopping and other amenities. Shortages are chronic, and prices, especially of housing, are high. These pressures, due mostly though not exclusively to the phenomenon, have begun to make life a little uncomfortable in Cambridge for the university, employers, and residents.

These shortages, high costs, and other pressures point to the urgent need for the Cambridge subregion to grow physically, in both the local and the national interests. Unless growth occurs, and the physical "holding capacity" of the area is enlarged, the risk is not only that existing businesses will not reach their full potential but that new ventures will not continue to be set up locally.

Such a conclusion about the need for growth is not universally shared. There is a long history of an antigrowth sentiment in the local community. This feeling results from a desire to keep Cambridge small and particularly its historic center unchanged and now an increasing feeling to keep the surrounding villages small and the countryside underdeveloped. (The last despite the fact that much of the countryside is flat and uninteresting and despite the impending release of land because of European agricultural surpluses.) Even among those who promote the need for physical development, there are differences in view as to what kind of growth should take place, how much, where and for whom.

Such stresses and strains are neither unexpected nor insoluble, but they will only be solved if all the big players—the local planning authorities, the university, and the high technology and other business communities, to name the most important—individually and collectively address the issues in a truly strategic sense. The initiative taken by the high technology firms in equivalent but much larger-scale circumstances in Silicon Valley in the late 1970s points to what can be achieved if there is the political will to do so.

It would be ideal to pretend that the issues are easy. Their local complexity is compounded by the growing national concern about the economic overheating that is occurring in the southern parts of Britain relative to the old industrial areas mostly in the north. Unrealistic as it may be, there is a sense that growth should be diverted away from places like Cambridge to disadvantaged areas. This is an issue that is becoming part of the national political agenda.

CONCLUSION

The Cambridge phenomenon has entered a fascinating and crucial period. It is still young and small in scale, and it needs to grow if it is to survive and prosper. Its continuing success will depend on many factors. This paper has highlighted two of the most critical: orderly growth of the subregion so that it remains an attractive place to live and work and so that the special character of Cambridge itself is retained, and maintaining the international excellence of the university.

It is salutary to observe that even in the evidently propitious circumstances of Cambridge, there is nothing automatic about sustaining successful high technology development over the long-term. To a great extent it is up to the participants, in their own and their mutual interests, to create the conditions in which this will happen.

In the past such conditions evolved and interacted spontaneously: Out of this the phenomenon spontaneously emerged, but now these very same factors are constraints to further development. To sustain growth of the phenomenon requires conscious and careful strategic thinking among all the key groups and, where their interests overlap, concerted action by them. The next stages of the phenomenon will indeed be challenging and will continue to offer food for thought about technology-based local economic development.

BIBLIOGRAPHY

Keeble, D. 1987. "Entrepreneurship, High Technology Industry and Regional Development in the United Kingdom: The Case of the Cambridge Phenomenon." Paper (with extensive bibliography) presented at a seminar on Technology and Territory: Innovation Diffusion in the Regional Experience of Europe and the USA. University of Naples, February.

Segal Quince Wicksteed. 1985. *The Cambridge Phenomenon: The Growth of High Technology Industry in a University Town.* Cambridge, England: Segal Quince Wicksteed.

———. 1986. "Strategic Planning Implications of High Technology Growth in the Cambridge Area." Cambridge, England: Segal Quince Wicksteed.

———. 1987. "University-Industry Research Links and Local Economic Development: Food for Thought from Cambridge and Elsewhere." Draft report (with extensive bibliography) for the Manpower Services Commission.

Chapter 6

SOPHIA-ANTIPOLIS AND THE MOVEMENT SOUTH IN EUROPE

Pierre Laffitte

The term *Sophia-Antipolis* has Greek origins. It stands for the city of wisdom, science, and technology. Sophia-Antipolis was founded in a nonurbanized area in Southern France near Nice in 1969 by a private, nonprofit company that I established to create a major research and science park in Europe. Three key factors have sustained the development of Sophia-Antipolis: the cross-fertilization of ideas, individuals, and organizations; a high quality of life; and close contact and interaction with the region and its constituencies.

In less than twenty years, Sophia-Antipolis has become the largest science park in Western Europe. About 150 large and medium-size companies from all over the world have set up operations there. Some like Dow Chemical, Dow Corning, Digital Equipment, IBM, Searle, and Cordis are U.S. companies. Others are Japanese, Swiss, English, and, of course, French. In the process about 6,000 direct jobs, and about 20,000 indirect jobs have been created.

SPIRIT OF ENTREPRENEURSHIP

In my opinion, however, the job creation from larger companies is not the most important achievement that Sophia-Antipolis has brought, and will continue to bring, to the French Riviera. The real impact lies in a new spirit of entrepreneurship that is evident in the region. Students, professors, and engineers are now looking to create new companies. The local elite, the chamber of commerce, and potential investors were accustomed to invest only in real estate. Now they are beginning to think in terms of creative capitalism by providing venture capital and seed money to emerging enterprises. Local government was interested in the experience from the beginning. However, it now plays an active role in the economic development of Sophia-Antipolis. The new motto of Côte d'Azur (Sea of Blue) is, French Riviera, the State of the New Enterprise.

STAGES OF GROWTH

Three stages of growth have marked the development of Sophia-Antipolis. The first stage focused on developing a science park in unusual surroundings. Sophia-Antipolis had little in common with other parks that were created around a major university or in the vicinity of preexisting industrial settlements. Without a university or industrial base, the first stage of development emphasized marketing a new high tech real estate location. In Europe, this is very rare. It could not have been achieved without a strong network of friends in academic, industrial, and political circles. Those friends formed a strong lobbying base and provided marketing support.

The second stage emphasized the process of cross-fertilization. People working in different corporations, in government research centers, and in universities do not easily mix, especially in Europe. Many ways have been tried, some of them successfully, to develop better interaction among and between these groups. In Sophia-Antipolis, cross-fertilization was encouraged through architectural planning and urban development. In other words, people needed pleasant and useful places to meet during the day. Restaurants, bistros, exhibition halls, like the old Greek Agora or the garden of Akademos and our Interaction Center Building. In addition, a critical component in this stage was a small team of men and women devoted to the task of developing this interaction.

The third stage is now beginning in Sophia-Antipolis. It involves helping turn ideas into entrepreneurial projects and then finding investors with seed money to fund the projects. In addition, sustaining emerging companies requires an environment conducive to entrepreneurship and an ability to provide sophisticated services, including patents and licenses, head-hunting, international marketing, financial options, electronic data bases and mailing, and other programs for small business. The Sophia-Antipolis Foundation and Technopole Service, a consulting organization, are helping this new development. This focus on entrepreneurship and new company development is the foundation for economic development in the park and the region.

CREATING AN INTERNATIONAL
TECHNOPOLIS NETWORK

Science parks and the technopoleis that are growing around them are increasingly trying to find ways to support emerging companies and thus spur economic development. To do this, they need help in building a network of support services and information on an international basis.

Three years ago, the International Association of Science Parks–Club des Technoples was created in Sophia-Antipolis to meet this need. This club has begun to act in three ways: sponsoring a series of national, regional, and international organizational meetings; distributing a trimestrial newsletter; and building data bases to help small, innovative companies. The Club des Technopoles is thus seeking to provide a way to disseminate information and network organizations, especially in countries where science parks and technopoleis are just beginning to develop.

DEVELOPING A EUROPEAN SUNBELT

Europe is now experiencing a great movement south, and Sophia-Antipolis is one indication of this movement. The European industrial revolution, which led to the world's industrialization, had most of its strength in the Hercynian backbone: Northern Europe from Wales to Bohemia, covering the coal fields of Great Britain, France, Benelux, and Germany, and the iron ore deposits of the same regions.

The Mediterranean sea, the sunny part of Europe and the region of antiquity with a long tradition of commerce, exchange, and cross-fertilization, has had the possibility to experience an economic comeback, by luring the men and women of high tech enterprises, like in California and the U.S. sunbelt states. Indeed, the first steps of this comeback are already visible in the Mediterranean region (see Figure 6–1).

Italy has had impressive success in modern economic development. Its economy has overcome that of Great Britain. Spain will certainly follow rapidly. The southern part of France is developing more rapidly than northern France. In Germany, there is more technology diversification and development in the south (Baden-Württemberg is the first high tech "Land" in the Federal Republic).

The major feature of modern Europe now is the active development of Lotharingia, with the Rhine/Rhône/Alps corridor from Alsace and Baden-Württemberg through Bale, Zurich, Lyon, Geneva, Grenoble, and Turin to

Figure 6–1.   Developing Technopoleis in Europe and the
Movement South.

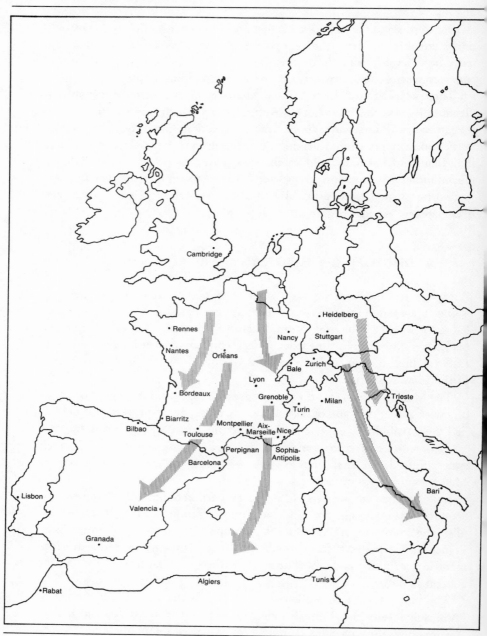

Aix-Marseille, Montpellier, Sophia-Antipolis, and Nice. This movement will rapidly develop westward toward Toulouse, Bordeaux, Biarritz, and Bilbao and southwest toward Montpellier, Perpignan, Barcelona, and Valencia.

In less than twenty years, the European sunbelt has become an attractive region for high technology development and venture capital. Perhaps even more significant, in my opinion, will be the future development of the North/South connection crossing the Mediterranean by developing technopoleis in southern Italy (Bari is already on the way), South Spain (Valencia and Granada), South Portugal (Lisbon), and Maghreb (Rabat, Algiers, and Tunis) (see Figure 6–1.) There are political as well as economic reasons for this movement. Considering the consequences of U.S./USSR military negotiations, Europeans must look to the survival of the European culture. This means, in my opinion, a European integrated army and the building of Eurafrica.

Europe with its increasing population needs more space. Its natural hinterland is Africa. As shown in Figure 6–1, the Rhine/Rhône/Alps and Milan/Turin/Sophia-Antipolis/Aix/Montpellier/Toulouse/Barcelona corridors are a capital core for future Eurafrica development. Building cooperative technopoleis with Europe and organizations in North Africa is one step that may help this long-term strategic purpose.

Part II

THE U.S. EXPERIENCE

Chapter 7

SILICON VALLEY
The Rise and Falling Off
of Entrepreneurial Fever

Judith K. Larsen and Everett M. Rogers

In early 1984 our book *Silicon Valley Fever: Growth of High-Technology Culture* was published. We described the rise of Silicon Valley in Northern California over the twenty-five-year period since 1960, stressing the spread of entrepreneurial fever as an essential ingredient in this spectacular success story. Exemplars of this entrepreneurial spirit in the microelectronics industry were Steve Jobs of Apple Computer; Adam Osborne of Osborne Computer; Nolan Bushnell of Atari, Chuck E. Cheese Pizza Time Theater, and Androbot (a robotics firm); Jerry Sanders of Advanced Micro Devices (AMD); and Bob Noyce of Intel. Where are these entrepreneurs today, three years later?

1. Steve Jobs was forced out of Apple Computer, the company that he cofounded, and has since started another firm.
2. Osborne Computer declared Chapter 11 in 1984 and has since reorganized, but without Adam Osborne, who left to form the Paperback Software Company.
3. Nolan Bushnell resigned from Atari, after having sold the videogame company to Warner Communications, who then managed the firm into financial disaster (it is now recovering, under different management). Pizza Time lost $75 million during a four-month period in 1983 and filed for protection from creditors under Chapter 11 in 1984. Androbot's first product was negotiated for sale to Atari in 1984, but the deal fell through. As happens in Greek tragedies, fate caught up with the hero (Betz 1987: 19).

4. Jerry Sanders's AMD, once the most profitable investment in the U.S. semiconductor industry, now languishes in the stock market, and Sanders has had to recant his once-proud boast that AMD would never lay off employees.

5. Bob Noyce's Intel, which we regarded in our 1984 book as the most innovative U.S. semiconductor firm, is now another of the many U.S. semiconductor firms struggling to stay alive in the face of Japanese competition.

What has happened to Silicon Valley? And to the thousands of millionaire entrepreneurs of its microelectronics industry? Has its entrepreneurial fever gone?

These questions are of crucial policy importance to the U.S. efforts to survive in the competitive international marketplace of high technology. Our purpose in this chapter is to explain the main events in Silicon Valley in the 1980s. We begin by briefly tracing the rise of the Silicon Valley technopolis. We hope that the present analysis of mature technopolis will provide some lessons learned for other technopoleis that have not yet approached maturity.

THE RISE OF SILICON VALLEY

Silicon Valley is an unusual community in several respects. It has immense riches, with an estimated 15,000 millionaires and at least two billionaires. It has a sunny, much-admired climate. It is the world's center for producing advanced information technologies: semiconductors, microcomputers, computer peripherals, and lasers. But Silicon Valley is particularly unusual because it demonstrates the role of entrepreneurial spirit in creating the new information society. Silicon Valley has been called the world capital of the information society (see Figure 7–1).

What is high technology? A *high tech industry* is characterized by (1) highly educated employees, many of whom are scientists and engineers, (2) a rapid rate of technological innovation, (3) a high ratio of research and development expenditures to sales (typically about 1:10), and (4) a worldwide market for its products. The technology is continuously changing at a rate much faster than in other industries. The main high tech industries today are electronics, aerospace, pharmaceuticals, instrumentation, and genetic engineering. Certain subindustries within electronics, like semiconductors and microcomputers, possess a technology that is advancing rapidly—so microelectronics is the highest of high tech.

Figure 7-1. U.S. Civilian Labor Force by Four Sectors, 1800–1980.

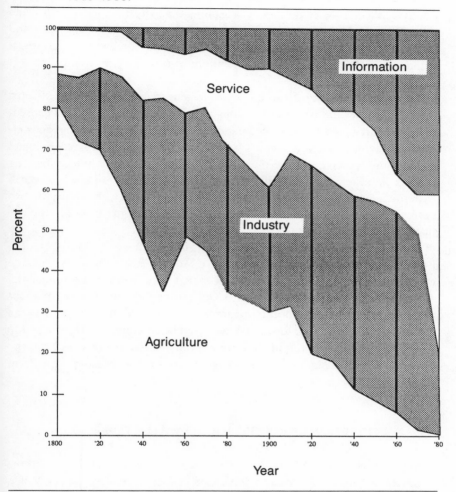

THE SPIRIT OF CAPITALISM

Max Weber's (1948) famous work, *The Protestant Ethic and the Spirit of Capitalism,* written about eighty years ago, argues that large-scale social changes result from a shift in basic social values rather than just from technological advances. Weber noted that the rise of Calvinism and Lutheranism in

Europe was associated with the growth of capitalism. The investment of capital in business enterprises by economic entrepreneurs was facilitated by a religious doctrine that considered business as a legitimate activity compatible with God's will. Although Weber's thesis has been much debated by social scientists and certain inconsistencies have been noted, nevertheless it represents a powerful explanation of particular historical developments in the Renaissance (Trevor-Roper 1972).

Many social observers feel that the present-day transformation of the United States and other industrialized nations into information societies is caused, at least in large part, by new information technologies, especially by the computer and by semiconductor chips. These advanced information technologies are certainly crucial, but we must ask why and how these innovations are created. A basic component in the current societal transition is what we call *entrepreneurial fever.* Entrepreneurship and information technology, the two critical factors in recent social change, are complexly interrelated in the rise of Silicon Valley.

Entrepreneurial spirit in Silicon Valley is analogous to the role of capitalistic entrepreneurship in the emergence of Western Europe's industrial society a century ago. In both cases, social change resulted from a fundamental entrepreneurial drive: The earlier case manifested changes in religious doctrine, and the contemporary example involves a culture moving from an industrial society to an information society. The economic entrepreneurship that supported the Protestant Reformation in the Renaissance and the rise of industrialism is similar to the entrepreneurial forces causing the current information revolution.

THE BEGINNINGS OF SILICON VALLEY FEVER

Just as Manchester, the Saar Valley, and Pittsburgh were once centers of the industrial society, today's information society has its heartland in Silicon Valley, a thirty- by ten-mile strip between San Francisco and San Jose (Figure 7–2). Silicon Valley is the nation's ninth-largest manufacturing center, with sales of over $40 billion annually. About 40,000 new jobs were created in the valley each year in the early 1980s. Its economy is among the fastest-growing and wealthiest in the United States.

Why did the microelectronics industry emerge in Silicon Valley? When and how did it begin? A small group of radio and electronics pioneers in the Palo Alto area included a young man named Frederick Terman, later to be called the "godfather of Silicon Valley" (Rogers and Larsen 1984). Terman grew up at

Figure 7-2. Location of Silicon Valley in Northern California.

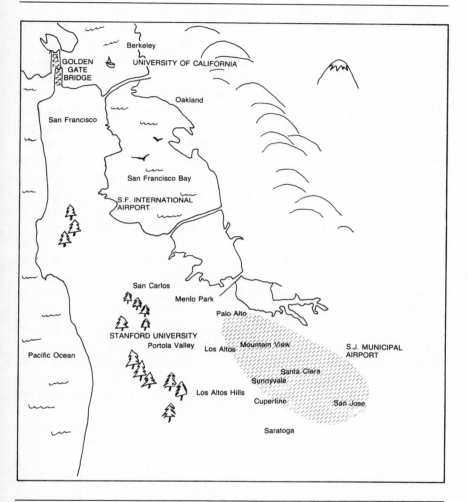

Stanford University, where his father was a psychology professor and where he later became an electrical engineering professor. Terman accepted a faculty position at MIT in 1924 but was stricken with tuberculosis and spent a year recuperating at his family home in Palo Alto. While there, he accepted a position at Stanford rather than at MIT because of the more healthful California climate. So it was somewhat of an accident that Terman located in what was later to be called Silicon Valley.

Terman favored close contact between the university and high technology firms, and he found jobs for his promising students in electronics companies, most of which were located in the East. Two of Terman's star students were William Hewlett and David Packard, who returned from the East Coast to complete a fifth year of study in electrical engineering at Stanford. The topic of Hewlett's master's thesis project, a variable-frequency oscillator, had a promising commercial application, so Terman advanced the two entrepreneurs a loan of $538 to start production in a garage behind their rooming house in Palo Alto. Today, both Hewlett and Packard are billionaires, and their company is the biggest success story in Silicon Valley. The Hewlett-Packard garage has changed little since 1938, when Bill and Dave moved their company to larger quarters.

Terman's most important contribution to Silicon Valley was his conception of a research park. His belief in the value of close university/industry ties led him to suggest leasing a large section of university-owned land to high technology companies. In 1951 the idea of a university industrial park was completely foreign, but Terman was a visionary. Hewlett-Packard and Varian Associates were among the first firms to locate in the research park. By 1955 seven companies were in the park, and by 1984 the space was filled with ninety tenants and 25,000 employees. Stanford Industrial Park was not only the first of its kind and the most successful, but it also has served as a model for scores of other high tech parks in the United States and abroad. The park also did a great deal for Stanford University, providing cash with which to hire top professors and thus improve the academic prestige of the university, which has risen considerably since 1960.

The other godfather of Silicon Valley is Dr. William Shockley, who launched the semiconductor industry. Shockley co-invented the transistor at Bell Labs in 1947. Silicon Valley probably would not be in California today had Palo Alto not been Shockley's home town. Shockley moved to Palo Alto, where his mother still lived, in 1955, to found Shockley Transistor Laboratory. It was later to become the direct or indirect source of most of the eighty semiconductor firms that started up in Silicon Valley.

Shockley launched his semiconductor company to develop commercially the transistor technology that he had invented at Bell Labs. He claimed that his goal was to make a million dollars (Rogers and Larsen 1984). But his company was short-lived and unsuccessful. Nevertheless, Shockley made a major contribution to the rise of Silicon Valley: He identified brilliant personnel. Among the eight bright young men whom he recruited to join his company were engineers and physicists who represented the cutting edge of semiconductor technology; most were from the east coast. The best known is Dr. Robert Noyce, who joined Shockley after leaving a position with Philco in Philadelphia.

The Shockley Eight were imbued with the entrepreneurial spirit that they learned from Shockley. Within a year, the eight enterpreneurs left Shockley to start a semiconductor manufacturing company of their own—Fairchild Semiconductor. The Shockley Eight then became the cadre of leaders for the semiconductor industry that sprouted in Silicon Valley. Fairchild Semiconductor was the spawning ground for scores of spin-offs.

The new companies usually were started with their founders' own capital and with loans from venture capitalists. Without the availability of venture capital, the proliferation of semiconductor firms in the 1960s and early 1970s could not have occurred. Even then, it cost big money to start a semiconductor company, much more than a couple of Fairchild engineers could save from their salaries. Venture capitalists knew that semiconductors were a growth industry, and a team of engineers with a bold new idea looked like a good proposition. Further, military customers usually were standing in line to purchase the new products. All the conditions were right for entrepreneurial growth. By 1983 eighty-five semiconductor firms had been launched in Silicon Valley, and it had become the U.S. center for this industry (Rogers and Larsen 1984).

Robert Noyce played a key role in spinning-off Fairchild from the Shockley Semiconductor Laboratory in 1957; in fact, he led the Shockley Eight in their defection. In 1968 Noyce left Fairchild to start another semiconductor firm, Intel, which later became the most innovative company in the semiconductor industry. It was at Intel in 1971 that Dr. Ted Hoff invented the microprocessor, a semiconductor chip containing the control functions of a computer (literally, a microprocessor is a computer on a chip). This miniaturization made possible the microcomputer and many other microelectronics products.

A critical mass of individuals and companies eventually developed in Silicon Valley. In 1976 Steve Jobs and Steve Wozniak built the first Apple microcomputer and launched their company, Apple Computer, Inc. Such role models attracted new entrants to the microelectronics industry. Silicon Valley's entrepreneurial opportunities were a beacon to like-minded individuals with a vision of starting their own firms. Once semiconductor firms were established in Silicon Valley, the area became the center for each of the new industries that applied semiconductors to a particular use: microcomputers, telecommunications, and video games, for example.

Silicon Valley is located in Santa Clara County, California because of Frederick Terman and William Shockley's associations with the county. Had Palo Alto not been their hometown, Silicon Valley might be located around Atlanta, Pasadena, or Keokuk. Table 7–1 describes the main inventions and events in the rise of the microelectronic industry.

Table 7-1. A Chronology of the Important Inventions, Events, and People in the Microelectronics High Technology Industry.

Year	Event
1912	Lee de Forest discovers the amplification qualities of the vacuum tube in Palo Alto, California, thus making possible radio, television, film, and other communication technologies.
1938	Hewlett-Packard is founded in a garage in Palo Alto by William Hewlett and David Packard, two of the first entrepreneurs in Silicon Valley.
1946	ENIAC, the first mainframe computer, with 18,000 vacuum tubes, is invented at the University of Pennsylvania.
1947	William Shockley, John Bardeen, and Walter Brattain invent the transistor at Bell Labs in Murray Hill, New Jersey. The transistor eventually replaces vacuum tubes.
1955	Shockley leaves Bell Labs to establish Shockley Semiconductor Laboratory in Palo Alto.
1956	Shockley, Bardeen, and Brattain win the Nobel prize in physics.
1957	The entrepreneurial spirit of Silicon Valley gets underway when Robert Noyce and seven other brilliant young engineers quit Shockley Semiconductor Laboratory to launch Fairchild Semiconductor. These cofounders later split off to launch over eighty semiconductor firms in Silicon Valley over the next thirty-five years.
1968	Noyce leaves Fairchild to start Intel.
1971	Invention of the microprocessor, a computer control unit on a semiconductor chip, by Ted Hoff of Intel.
	Silicon Valley is named by the late Don Hoefler, then editor of a local electronics newsletter.
	Nolan Bushnell designs Pong, launches Atari, and the videogame industry is begun.
1976	Steve Jobs and Steve Wozniak build the Apple microcomputer.
1980	Apple goes public: Art Rock, the venture capitalist who had invested $57,000, earns $14 million; Jobs is worth $165 million.
1982	About 3,100 microelectronics firms exist in Silicon Valley; two-thirds have less than 10 employees, and only fifty or so have more than 1,000 workers.
1984	Silicon Valley has 15,000 millionaires and 2 billionaires.

ENTREPRENEURIAL SPIRIT IN SILICON VALLEY

Silicon Valley companies are constantly starting up, growing, merging, being acquired, or fading away, making it difficult to know exactly how many firms exist at any one time. A careful count in 1982 identified 3,100 electronics manufacturing firms in Silicon Valley. In addition, companies supporting the electronics manufacturers, such as firms engaged in marketing, advertising, research and development, consulting, training, venture capital, legal, and other support services brought the total number of firms in the Silicon Valley electronics industry to about 6,000. Another 2,000 companies are in nonelectronics high technology fields such as chemicals, pharmaceuticals, and biotechnology. So the total number of high tech firms in Silicon Valley was about 8,000 in 1983.

Most of these firms are small. Over two-thirds have fewer than ten employees; 85 percent have fewer than fifty. Public attention is concentrated on the fifty-four electronics firms with more than 1,000 employees, companies such as Hewlett-Packard, Intel, and Apple Computer. These giants constitute about half of the total work force. But thinking about the Silicon Valley technopolis, one should remember that the great majority of the firms are very small.

The entry ticket to the microelectronics industry is a new product. It is the promise of a new product that helps attract venture capital to start the company. Then the new company's growth is fueled by sales of the product and by a strategy of continuous technological innovation. In order to compete effectively, the company must outgrow other companies producing the same product. Such rapid expansion puts great strain on a new firm, as it seeks to navigate this "adolescent transition" from being a small, informal group to becoming a large bureaucracy. The number of employees may double annually for the first years of the firm's existence. People no longer know each other by their first names. Then they begin to ask, "Why isn't it any fun around here anymore?" Ultimately, the new firm must change from an entrepreneurial orientation to putting emphasis on management, as it becomes an established bureaucracy.

So while the entrepreneurial spirit is directly involved in creating technological innovations and in the start-up of new firms around these innovations, entrepreneurial fever does not fit well with managing a fast-growth or large and stable firm. The adolescent transition poses a threat to the survival of a high technology firm. Sometimes an entrepreneur learns to become a more bureaurcratic manager and thus leads the firm through the transition. In other companies, the founding entrepreneur is moved up to a figurehead position like chairman of the board, while a professional administrator is brought in to operate the company. Often, the entrepreneur leaves the firm, perhaps to launch yet another company. Steve Jobs is an example.

KEY FACTORS IN SILICON VALLEY'S RISE

What factors were essential in the rise of Silicon Valley?

1. *Availability of technical expertise.* Silicon Valley high tech companies depend on individuals who can design semiconductor cleanrooms, tool delicate fixtures, and design innovative products. Few other places in the world can match the pool of experienced, specialized high tech brainpower of Silicon Valley. Companies that need access to such expertise have little choice but to locate where these intellectual resources are concentrated, creating a further agglomeration of microelectronics firms.
2. *Infrastructure.* A new firm is dependent on suppliers, financiers, markets, and other infrastructure. Start-ups are more likely to occur, and more likely to succeed, where the necessary infrastructure exists.
3. *Venture capital.* Silicon Valley is a prime center of venture capital: Over one-third of the nation's venture capital companies have offices in Silicon Valley. Venture capital is money placed in new or young high technology companies with a potential for rapid growth. A venture capital firm serves as an intermediary between investors looking for high returns for their money and entrepreneurs in search of needed capital for their start-ups. Venture capital firms invest their money largely on the basis of the potential value of an entrepreneur's idea, a collateral that conventional bankers consider worthless. Entrepreneurs give up a percentage of the ownership of their new company, often about 50 percent, in exchange for acquiring capital.
4. *Job mobility.* Annual job turnover in Silicon Valley has been about 30 percent. In other words, the average employee would have three different jobs in ten years. Such turnover is encouraged by a shortage of qualified, experienced personnel. To counter such rapid turnover companies offer short-term benefits and incentives such as stock options, recreational facilities, and training programs to their employees.
5. *Information-exchange networks.* Silicon Valley is not just a geographical place or simply the center of the microelectronics industry; it is a network of networks. Extensive personal contacts facilitate information exchange. News about people changing jobs, about new products, about manufacturing successes and failures, all are instantly common knowledge.
6. *Learning entrepreneurial fever from local role models.* The spin-off of firms in Silicon Valley represents a modification in industrial structure, which stems from the basic social change of the United States to an information society. The rate of spin-offs is the key indicator of the rise (and fall) of a high technology complex.

Successful entrepreneurs became the superelites of Silicon Valley, something like Hollywood stars or pro athletes. Silicon Valley's star entrepreneurs receive much attention from the mass media. They serve as much-admired role models, adding further to the entrepreneurial head of steam in Silicon Valley.

Entrepreneurship is best learned by example. When an individual learns of successful role models like Bill Hewlett, David Packard, Bob Noyce, Steve Jobs, and others, they begin to think, "If he did it, why can't I?"

Silicon Valley represents a special kind of information society heartland based on continuous technological innovation, vigorous competition, and entrepreneurial spirit. High technology culture spawned millionaires at a remarkable rate. It contributed new jobs and tax dollars to the local area. High tech industry also was good for the United States, making it the world center of the emerging information society, providing an economic bright spot in an otherwise dreary picture of smokestack industries and displaced workers. Silicon Valley was not planned by a central authority, and local, state, and national governments did not play a very important role in its development.

That *was* the picture of Silicon Valley. What happened next?

CHANGES IN SILICON VALLEY IN THE 1980s

The feverish growth of Silicon Valley that caught public attention in the early 1980s has quieted down in recent years. The slow-down leads one to ask whether Silicon Valley is dead. Is its high tech entrepreneurship a thing of the past?

As we stated previously, Silicon Valley is based on two basic technologies: semiconductors and computers. For some years, popular attention focused on certain large companies in the two industries organized around these technologies: Examples are Apple Computer, Atari, Advanced Micro Devices, and Intel. The large companies, especially firms that offer consumer-oriented products, have received most of the mass media attention. But the backbone of Silicon Valley always has been its thousands of small technology-based companies. These companies are not household names. Yet they form the foundation of Silicon Valley's entrepreneurial high tech industry.

Silicon Valley's larger companies have indeed changed their operations in recent years. As the big companies reorganize, they change the face of Silicon Valley, at least the face seen by the public. But we maintain that the underlying entrepreneurial fever, with its enticement of fortunes to be made, remains rather firmly in place in Silicon Valley.

As high tech business turned down, U.S. semiconductor firms migrated overseas, most notably to Asia. When this exodus began to be noticed, the initial reaction was one of surprise and shock. *Business Week* (Wilson 1985) has run a cover headline, "America's High-Tech Crisis: Why Silicon Valley Is Losing Its Edge."

Indeed, Silicon Valley has changed dramatically in the 1980s. Foreign trade and increased international competition strongly affect the semiconductor industry. The high tech trade problem of the United States centers largely on a growing deficit with Japan. The trade deficit from the exchange of electronic products between the United States and Japan grew from $9 billion in 1983 to $15 billion in 1984, and doubled again in 1985 (Wilson 1985). Until 1982 electronics represented a trade *gain,* rather than a trade *deficit,* with Japan.

Other factors also influenced large chip-makers to leave Silicon Valley. One is cost. As land prices continue to rise in Silicon Valley, high tech companies look for other locations to find cheaper land, lower labor costs, cheaper electrical power, and less expensive materials.

Diversification is another major incentive for corporations to move. As Intel's James Jarrett said, "You don't want to have all your eggs in one basket. We believe in spreading things around, so you're not necessarily subject to one type of problem in one area." Natural disasters, especially earthquakes, in Silicon Valley, are one of the semiconductor industry's concerns.

U.S. semiconductor firms also seek locations closer to major customers. The potential market for many U.S. companies is split between the United States and Asia, especially Japan and Korea. To be competitive in a market requires that firms be located near their customers. Direct contact is necessary to establish accurate communication and to maintain network relationships. Customer confidence is also generated by a strong and visible presence. As Roger McDaniel of Monsanto Electronic Materials explained, "Putting a multimillion dollar investment in Japan shows our commitment and determination to succeed there."

Microelectronics companies also want a global presence so they can be responsive to the needs of their clients. *Fortune* 500 multinationals want the ability to deal with one company anywhere in the world. So semiconductor companies must have their facilities and representatives worldwide.

Quality of life is another factor behind the geographical spread of Silicon Valley microelectronics corporations, although it is one that is not easily quantified. Quality-of-life components have deteriorated. Silicon Valley traffic is a mess. Dense development of industry and housing has created urban congestion. Housing in Santa Clara County is among the most costly in the nation.

The long hours at work limit people's involvement in social activities. Such a deteriorating quality-of-life makes it difficult for Silicon Valley to attract workers, especially workers at lower pay levels.

In spite of these ominous developments, the move of microelectronics manufacturing from Silicon Valley may not be as serious as it first appears to be. Following the *Business Week* story heralding the "end" of Silicon Valley, *Forbes* ("Custom Chips" 1985) pointed out that more than sixty semiconductor companies have sprung up since 1979, the year the Japanese began making inroads into the U.S. microelectronics industry. Among these newer companies are such strong contenders as LSI Logic, Cypress Semiconductor, VLSI Technology, Sierra Semiconductor, and Integrated Device Technology.

The difference between the newcomers and the larger, established semiconductor companies is their product line. The established companies depended on memory chips, products that became commodity items. The Japanese realized the potential pervasiveness of commodity electronics components and invested millions of dollars to produce memory chips. Clearly, they have succeeded. The newer U.S. companies have ignored memory chips, focusing instead on application specific integrated circuits (ASICs). ASICs are becoming big business, and the Japanese firms now want to move into this market too. But high value-added products such as ASICs rely on other factors than production skill to influence sales. These other factors include close relationships with cusotmers and state-of-the-art capacity. In these factors, the newer U.S. semiconductor companies uniquely fill a market niche.

The move of semiconductor manufacturing activities out of Silicon Valley was underway even in the days of Silicon Valley's greatest growth. Until the mid-1970s Silicon Valley semiconductor chip companies carried out the overwhelming majority of their manufacturing locally. It was this activity that gave Silicon Valley its name. However, by 1981 only half of the production capacity of semiconductor manufacturers remained in Silicon Valley. By 1987 that figure is down to only one-third. It will continue to shrink.

Semiconductor manufacturers use several different strategies in moving their operations out of Silicon Valley. Some companies simply transfer their production plants to other sites, as AMD moved its manufacturing plants from Silicon Valley to Texas. Another method finds companies keeping their existing manufacturing operations in Silicon Valley but choosing other locations for expansion or modernization. Finally, some new semiconductor companies avoid Silicon Valley entirely and set up production operations elsewhere. Only their company headquarters and R&D division are in Silicon Valley.

Currently, nearly 90 percent of AMD's production is at new plants in Texas. The company has two wafer fabrication facilities remaining in Silicon Valley, compared to eight wafer fabs five years ago. Of the two remaining Silicon Valley wafer fabs, one does only limited manufacturing.

Despite the sharp decrease in semiconductor chip manufacturing, no one is suggesting that Silicon Valley will give up its title as the center of the U.S. semiconductor industry in the near future. Semiconductor manufacturing continues to move away, but Silicon Valley is more firmly devoted to advanced microelectronics R&D, to applications start-ups, and to headquarters administration. New semiconductor companies continue to sprout; 60 percent of all semiconductor start-ups since 1978 are located in Silicon Valley (Schmitt 1987). Further, the number of Silicon Valley electronics jobs continues to grow.

Employment patterns have changed in predictable patterns with the shift from emphasizing semiconductor manufacturing to research and development. With decreased semiconductor manufacturing, the number of semiconductor manufacturing employees in Silicon Valley has shrunk. In 1981 nearly 55 percent of Silicon Valley firms' wafer fab employees worked in Silicon Valley; in 1987 that figure has fallen to only 38 percent (Schmitt 1987). Further, about 75 percent of all new wafer fab jobs that Silicon Valley chip-makers have added in the past five years have been located outside the valley. In spite of this decrease in wafer fab jobs, the total number of Silicon Valley electronics employees has more than *doubled* since 1981 as a result of overall growth in the microelectronics industry (Schmitt 1987).

THE FUTURE OF SILICON VALLEY

What, then, do these changes mean? Silicon Valley's trend away from manufacturing may represent a face of the future. Many of the newer semiconductor companies in Silicon Valley do not intend to be involved in manufacturing. They design and market chips but contract with other companies, usually Japanese or Korean companies, that actually make the product. The Silicon Valley companies do the design, marketing, applications, and value-added work. These are the key information worker functions in a high technology industry, in an information society.

Silicon Valley's semiconductor companies produce the basic technological products supporting information-processing and transmission. These companies know how to use the information society to their advantage. For the most part, semiconductor companies and microcomputer companies have been able to manage their far-flung operations, so that the effect of geographical location is minimized. These companies have excellent communications systems, often utilizing satellite telecommunications to link their distant operations.

They have already moved into the future, in a communication sense. They substitute information for transportation.

The trends reviewed here that influence Silicon Valley's microelectronics industry, and the rest of the U.S. electronics industry, are not isolated phenomena. Each factor influences another. The future will see greater reliance on application specific integrated circuits, the semiconductor chips in which the United States is presently strongest. Advances in software and CAD technology are also critical in transforming Silicon Valley's microelectronics industry into a more service-oriented business and one in which skilled manual labor is less important. The importance of U.S. attention to value-added products is underscored by the shift in manufacturing to Asia, which will further add to Japanese competition with the United States in the value-added market.

To successfully compete, Silicon Valley companies must focus on identifying market niches and in specific applications of microelectronics technology. They must stress customer relations and service over internal manufacturing capabilities, thus emphasizing the gap between those companies that manufacture and those that supply (Boss 1986).

Silicon Valley is going through a period of major restructuring. The valley is decreasing its reliance on its two basic industries—semiconductors and computers—and expanding its focus to encompass the *applications* of microelectronics components. This move to applications brings a much wider array of industries and activities into Silicon Valley. The discontinuities of the major transition in Silicon Valley's structure is a major reason for the high rate of lay-offs in the 1980s. However, most individuals who are laid off in Silicon Valley find another job within a few months. Our research shows that these individuals pay a high price, psychologically, for being laid off (because work is so central to professionals in Silicon Valley). Mental problems, divorce, and other signs of stress characterize the laid-off employee. But we know of Silicon Valley firms that are hiring new employees (with certain qualifications) on the same day that they are laying off hundreds of other employees. Thus we see the nature of the transition now occurring in Silicon Valley and its human cost.

Silicon Valley is now in its second phase. The first spurt of growth from 1960 to 1980 is over. The attention generated by this expansion and integration will produce a shift away from larger, centralized corporations to smaller, decentralized companies. Whereas Silicon Valley's childhood years addressed semiconductor and computer manufacturing, the adolescent years have seen semiconductor chips and computers being *used* by other industries. New subindustries have emerged as a result of microelectronics applications: medical electronics, communications systems, automotive electronics, telecommunications, biotechnology, and others. Instead of two main industries—semiconductors and computers—Silicon Valley now serves as headquarters for a broad base of microelectronics applications industries.

Ironically, the number of microelectronics-based industries in Silicon Valley continues to grow, but most of these firms do not produce consumer products. They are not very visible to the public. The current transformation underway in Silicon Valley is visible to veteran Silicon Valley watchers, but it is almost entirely ignored by the popular press.

Silicon Valley has not lost its entrepreneurs; they have simply moved to new, more profitable arenas. Present-day entrepreneurs form strategic alliances, especially with Asian companies, thereby bringing together the knowledge of the applications experts with the know-how of the manufacturers. Such collaborative alliances will produce the profits of the future by providing services and specialized products that the large, multinational companies will not be able to address.

A basic value change has been occurring in Silicon Valley in the 1980s: from strict competition to certain forms of collaboration. Much competition continues to exist, of course, as firms in the same industry seek to outperform each other. But today these competing companies are also likely to share a technology licensing agreement, to be costockholders in a university-based R&D consortium (like the MCC in Austin or the Center for Integrated Systems at Stanford University), and to join in other relationships in which neither party has complete control over the others. Much of the stress on collaboration in the microelectronics industry in the 1980s is due to the threat of international competition, especially from Japan. Silicon Valley is beginning to display certain characteristics of the M-form society, described by Ouchi (1984).

Thus we conclude that the temperature of Silicon Valley's entrepreneurial fever is cooling off somewhat in the 1980s. But the microelectronics industry of Santa Clara County is still viable and, in fact, continues to grow. This industry is in a major transition today, but these changes stem from the maturity of the Silicon Valley technopolis, not from any inherent weaknesses.

BIBLIOGRAPHY

Betz, Frederick. 1987. *Managing Technology: Competing through New Ventures, Innovation, and Corporate Research.* Englewood Cliffs, N.J.: Prentice-Hall.
Boss, Michael. 1985. "Semiconductor Downturn at Home Spawns Migration." *Corporate Times* (June).
———. 1986. "The Dataquest Semiconductor Megatrends." *Research Newsletter* (October).

"Custom Chips: The New Look in the Semiconductor Industry." 1985. *Forbes* (June 17).

Ouchi, William G. 1984. *The M-Form Society: How American Teamwork Can Recapture the Competitive Edge.* Reading, MA: Addison-Wesley.

Rogers, Everett M., and Judith K. Larsen. 1984. *Silicon Valley Fever: Growth of High-Technology Culture.* New York: Basic Books.

Schmitt, C.H. 1987. "Chip Firms Flight Changes the Face of Silicon Valley." *San Jose Mercury News* (February 1).

Trevor-Roper, H.R. 1972. *Religion, the Reformation and Social Change, and Other Essays.* London: Macmillan.

Weber, Max. 1948. *The Protestant Ethic and the Spirit of Capitalism.* New York: Scribner's.

Wilson, J.H. 1985. "America's High Tech Crisis: Why Silicon Valley Is Losing Its Edge." *Business Week* (March 11): 56–67.

Chapter 8

ROUTE 128
Its History and Destiny
James W. Botkin

Today Route 128 is synonymous with high tech success. It wasn't always. And it won't always be. Many other states would like to emulate the success of Route 128. Regional economic development officials, chambers of commerce, and far-sighted businesspeople are envious of the Massachusetts cornucopia of electronics companies, the growing biotech research, and garage-based advanced materials companies—not to mention the lowest joblessness rate in the country and a governor flirting with the 1988 presidency based on a state tax cut and a budget surplus.

Instant envy is natural. Instant replication is impossible. And just as forty years ago Route 128 was a disaster area, so forty years from now the smart ones will be those who ignored today's Route 128 and learned from its checkered history.

ROUTE 128: A HISTORICAL PERSPECTIVE

In a nutshell, Route 128's history is seven lean years for every seven fat ones. This technopolis is as close to a free market economy as possible. It has abandoned the three Rs: no regulations, no rules, no restrictions. Instead, the area builds on unlimited brainpower, abundant venture capital, and an endlessly proud blue-collar labor force. This is the stuff of Yankee ingenuity, the kind that Max Weber wrote classics about. It is also the kind that another economist wrote about. Far more significant than Weber's (1948) *The Protestant Ethic and the Spirit of Capitalism* was Kondratieff's (1935) *Long Wave Economic Cycles.* That is Route 128—cycles: long periods of success equally punctuated by full stops of depression.

Compared to other parts of the country (Texas excepted), Route 128's cycles are more severe than most. That is, the crest of the wave is higher (higher today, say, than Silicon Valley—just compare mighty DEC with rainbow/ephemeral Apple). The depth of the depressions are also lower than most anywhere else (compare the United Shoe Machinery plant on Route 128 to RJ Reynolds tobacco in North Carolina). Texas is excepted, because wild swings in the price of petroleum produce joy and misery even more quickly and extremely than its Yankee counterpart can.

Not only are the amplitudes of the New England swings extreme, but their regular pattern is accelerating. In other words, the joys and miseries are both shorter-lived. New England's ice economy, for example, lasted from 1620 to nearly 1870, whaling and the China trade from 1720 to 1880, and textiles and shoes from 1840 to 1940. High tech has thrived for about thirty years—if dated from 1957, which is the year Ken Olson founded Digital Equipment Corporation in an old textile mill by a pond used for garnering ice by a previous generation. Contrary to the neatness of theory where the new economic activity grows on the site of the old, few whales have been harpooned in the Old Mill on Maynard Pond (although it has been rumored that the hapless DEC engineers who designed the company's ill-fated Rainbow PC were made to walk the plank.)

HIGH TECH PEAK?

High tech has been around for thirty years. DEC and the region are at an all time high, whereas Wang (the other great garage story of the area) is on the way down. However, a collapse of the local high tech industry is predicted by Jay Forrester, the MIT professor who invented core memory in 1947, from which MIT made millions in royalties. Forrester is the perfect Yankee Brahmin professor. Born in Nebraska, he was imported from Cambridge University, England, during World War II. He stayed on at MIT and built the famous Systems Dynamics computer models. In 1980 Forrester successfully predicted the collapse of farmland prices in his native Nebraska.

Now Forrester says that Route 128, as we know it, is finished. "As we know it" is key: It means electronics. The governor of Massachusetts, Michael Dukakis, introduced a bill to support biotechnology, photovoltaics, advanced polymers, deep sea mining, and other alternatives to high tech electronics.

Dukakis's plan is to generate Kondratieff's new wave before the old one breaks. Few people believe him and Forrester—so few that the governor could not fund the new programs. But at least they are there on paper for historians to record.

Another reason that Dukakis could not fund the alternative programs has to do with another Massachusetts peculiarity. The new programs fell into the lap of the Massachusetts lieutenant governor. In Massachusetts, like many U.S. States, the governor and his deputy are rivals. They fight rather than support one another.

RENAMING ROUTE 128

Route 128 is called Route One-Two-Eight by the Europeans and Arabs eager to invest in it. The Asians—Japanese and Chinese use pictograms to describe the area—call it "the ring." In New England, Route 128 has two names, which reflect both its history and its future. The first is Yankee Division Highway, which reflects its past. It was so named by the Veterans of Foreign Wars in honor of the men killed in the Spanish-American War and World War I. Route 128 was then at its peak in textiles and provided military uniforms and leather boots for soldiers. Leather boots were put on the strategic materials list in order to block imports that might threaten United Shoe Machinery. (This reminds me of the current concern with protecting semiconductor chip manufacturing in the United States.)

Ride along Route 128 today, and at Exit 26 you will see a small white sign that says *Yankee Division Highway*. Next you will see a large blue sign that says *America's Technology Region*. And if you get closer, you will see that the word *region* is imprinted on one of those paste-over signs.

What the paste-on *region* is covering is the word *highway*. And the reason is classic. In 1980 the high tech community wanted to honor America's Technology Highway. The VFW, with a longer view of history, protested. Since 1916 Route 128 has been the Yankee Division Highway, not America's Technology Highway. One cannot have it both ways, the VFW said. So the compromise by the high tech community was to call the area America's Technology Region.

THE ICE INDUSTRY

In 1640 Fresh Pond in Cambridge was the center of an emerging ice business. Enterprising young men, fresh out of Harvard (founded 1636), had learned to apply technology—the two-man push-pull saw—to cutting blocks of ice that were loaded onto horse-drawn sledges and carried to markets as distant as fifty miles. Business boomed, especially with the standardization of ice block sizes to maximize the sled load.

Then R&D took over. The opportunity was the big markets of overheated Spanish conquistadores in Florida. The problem was that ice melted. The solution, funded by venture capitalists at the time working with scientists and engineers, was sawdust from the emerging lumber industry. Sawdust sufficiently retards the melting of ice so that a fast-sailing ship can carry a load of ice from Salem, Massachusetts, to Sarasota, Florida. Thus were born the Yankee clipper ships.

The Cambridge economy boomed and so did that of surrounding areas with ponds. Route 128 passes no fewer than fifty ponds on its 100-mile route. Its connecting roads also boast rotaries that were made for horse traffic so that a horse with blinders can turn around without distracting the other horses because one can't see the other. Clever! And a good way to get ice to market without grid lock, or ice jam.

In 1860 a man in Florida (probably a descendant of one of those warm Spanish conquistadores) invented something that came to be known as refrigeration. This sounded the death knell of New England's ice industry. Representatives of the Church of New England, who also served on the then-equivalent chamber of commerce, protested. "Refrigeration is the work of the devil," they declared. They succeeded in blocking the use of iceboxes and especially refrigerated railroad cars for many years, but theirs was a losing battle, and Route 128 lost its ice industry to such an extent that little evidence of its existence remains today—except the historical record.

THE WHALING ECONOMY

In the late eighteenth and early nineteenth centuries, whaling replaced ice as the region's driving economic force. Although not centered strictly on Route 128, which is virtually landlocked, New Bedford and Nantucket became the centers of whaling ships. The economic effect of the whaling economy was far-reaching, however. It stretched all the way to China. Trade with mainland China was brisker than today, even though a single voyage consumed three to five years (no Panama Canal yet). A barrel of South Pacific crude was worth more in 1787 than its OPEC equivalent today. And a Massachusetts ship filled with whale oil could light 35,000 homes for a year.

The whaling trade brought boom times to Route 128, which housed the distribution points for whale oil. Its factories made the lamps, refining equipment, and all the Industrial Revolution capital goods that accompanied this source of energy. But whaling died out because, as overeager men will do, they killed too many whales. By 1845 a Nantucket ship had to search the south Pacific for three years and would often return home half empty, compared to three-month voyages and full cargoes fifteen years earlier. Nantucket went into a century-long depression, later to emerge on the waves of tourism. New Bedford also went into a long depression and has yet to emerge. Route 128 lost all its whale-oil-intensive industry and desperately searched for a substitute.

TEXTILES AND SHOES

The successor to the whaling-based economy was weaving. In 1821 Francis Cabot Lowell built a textile mill on the Merrimac River, not far from Route 128. From visits to England, Lowell brought to Massachusetts the latest technology for cotton spinning and weaving. In a way reminiscent of contemporary perceptions of Japan as a nation of copiers, Lowell adapted and refined British ideas by converting their technology into one continuous process from raw cotton to woven fabric. By 1900 Massachusetts was the world's largest producer of woven fabrics.

Eighty years later, textile production in New England was dead. So were shoes, which accompanied textiles. The United Shoe Machinery (USM) story is an interesting one. In 1902 USM organized the world's first industrial R&D facility at its Beverly plant just off Route 128. The company produced shoe-making machinery. Its greatest innovation was not technological but marketing: It leased its machines and refused to sell them. "Mainframes" leased from USM led to one of the country's greatest corporations and also to a famous antitrust suit. Another three-letter company learned this lesson well. Fifty years later, IBM followed USM's strategy to the letter, except with more political savvy. Instead of fighting antitrust suits, it chose to fight AT&T. However, IBM is based in the state of New York, which does not have a Route 128 or any technopolis, for that matter.

In 1980 USM went the way of the ice and whaling industries. The company was acquired by the Connecticut-based Emhart Corporation, a conglomerate that manages the manufacturing of machinery to make glass bottles, locks, rivets, and shoes. Recently USM announced the lay-off of another 100 workers, leaving the factory largely an empty shell.

DOCUMENTING THE DEPRESSION OF ROUTE 128

The famous book about the transition from USM to IBM, or the continuing saga of Route 128, is *The De-Industrialization of America* by Bluestone and

Harrison (1982). These authors focused on the collapse of the textile industry and claimed that the high tech economy did nothing to ease the suffering of the transition. This ran counter to the prevailing mood of the time, which was bullish on high tech. Many publicly repudiated Bluestone and Harrison even though they produced statistics that showed that less than 1 percent of the old labor force ever found a job in the new high tech labor force! Just because they were right did not mean high tech was a bad guy (which they claimed, and I still object to that).

But what Bluestone and Harrison document is the pain and suffering of Route 128 depression times. Entire communities such as Lowell, Massachusetts, were decimated. And such decimation was not limited to the textile industry. Waltham, Massachusetts, still has one of the largest clock and watch factories in the United States. Located on the Charles River, in the company's heyday it successfully challenged Swiss production with cheap U.S. watches. In more recent times, it had to close operations because of Japanese competition using cheap Asian watches. The factory windows are now bricked over, and its machinery is rusting.

LESSONS LEARNED

There are three main lessons to be drawn from the Route 128 experience. First, money and brains drive the economy. The businesses come and go, but the financial institutions, banks, and universities remain. The Boston Ice Company, New Bedford Oil Supply, Waltham Watch, and even the mighty United Shoe Machinery have either died or lie paralyzed. Yet Harvard, Boston University, MIT, and Worcester Polytech still thrive. State Street Bank, Shawmut Bank, and Cabot, Cabot, and Forbes are fine old financial institutions that outlasted each of the industrial cycles. New financial institutions, such as Fidelity Investment Company, are taking the country by storm. In the midst of technopolis, money-opolis and university-opolis are the stabilizing *sine qua nons* of the region.

Second, money and brains are built slowly and painstakingly. They do not come from incubators, incentives, or other government policies. This is not entirely true, of course. Government policy can play a major role. MIT was founded as a land-grant institution and to this day still receives land grant money ($23,000 in 1982). Government regulation of the banking industry has held banks in check, concentrating them in Boston, for example. Interestingly, the movement to subvent outmoded banking laws is what gave rise to the Fidelity Company, which itself chose to locate in Boston.

But very few incentives have led directly to the establishment of technopoleis. Where incentives are useful is in the slow building of the conditions where a technopolis might emerge. But emergence through giverment (read *government*) incentive was never the case. The one exception was military contracts that helped Jay Forrester and the chip industry. Massachusetts today receives an enormous share of defense-dollar research. MIT gets twice as much as any other major university, with the possible exception of Johns Hopkins in Baltimore.

Third, regions that seek to attract high tech and build a technopolis should remember that

1. It takes a long time to accumulate history. The nearest equivalent is North Carolina's Research Triangle Park, and they have been at it for over twenty-five years, nearly since the inception of the high tech revolution. North Carolina is still pretty small compared to the size of Silicon Valley and Route 128.
2. Government polices and incentives are more needed in the depression part of the cycle than in the inception point. They are probably more effective then, too, because it is easier for government to soothe people than to stimulate them.
3. High tech is no more stable or unstable than ice, whaling, or shoes. Of the three, it is actually closer to ice because it neither kills animals nor chains workers to machines. Then again, high tech does not have much of a history yet.

BIBLIOGRAPHY

Bluestone, Barry, and Bennett Harrison. 1982. *The Deindustrialization of America: Plant Closings, Community Abandonment, and the Dismantling of Basic Industry.* New York: Basic Books, Inc.

Kondratieff, N.D. 1935."The Long Waves in Economic Life." *Review of Economic Statistics.* 17 (November).

Weber, Max. 1948. *The Protestant Ethic and the Spirit of Capitalism.* New York: Scribner's.

Chapter 9

THE ROLE OF RENSSELAER POLYTECHNIC INSTITUTE Technopolis Development in a Mature Industrial Area

Pier A. Abetti, Christopher W. Le Maistre, and Michael H. Wacholder

MYTH AND REALITY IN THE CREATION OF TECHNOPOLEIS

The myth of technopolis, an utopian radiant city based on science and technology, is essentially a creation of the humanistic mind of the renaissance, perfected during the era of illuminism. Philosophers, political planners, and social prophets have for generations been intrigued by the idea of creating a perfect "city of light," designed and ruled by wise scientists, where research and innovation are a way of life and invention and creativity are venerated as the highest of virtues. These ideal cities ($\pi o \lambda \xi$ = city) are the poles ($\pi \jmath \wedge \epsilon \omega$ = to turn around, to offer to the market) around which the economy of the nation will grow and its society will progress.

Thus, dreams have permeated scientific thought and visions have been drawn from political reflection. Such utopias appear as plans for evading beyond history, as refuges against recurring crises, as "shell-spaces," closed to economic chaos (de Kerorguen 1985: 234–38). Yet these cultural projects represent aspirations toward a desirable new order and, like all myths, continue to haunt our subconscious to this day. In fact, several modern technopoleis have been given mythical or classical names: Sophia-Antipolis, Rennes-Atalante, and Bordeaux-Technopolis in France; Technopolis Novus Ortus in Italy; Mechatronics Valley and Computopolis (Tama New Town) in Japan; Akademgorodok in the USSR, and so forth.

125

Probably the first example of a utopian technopolis is the *City of the Sun* of Campanella (1623)—a polyvalent and polytechnical city, a showcase of science, an immense institute of continuing education, based on emulation and exaltation of inventors. Here science is not reserved for an aristocratic scientific elite (as in the *New Atlantis* of Francis Bacon, 1627) but instead is the basis for egalitarian training. The City of the Sun is ruled by a priest, Hoh, the Methaphysic. Under him are three chiefs: Pon (Power), Sir (Wisdom), and Mor (Love) (Campanella 1623):

> Wisdom is charged with the direction of the liberal and mechanical arts, and of all sciences, with their respective officials, doctors and schools. . . . There is a single book called *Knowledge*, in which all sciences are inscribed with marvellous conciseness and clarity. This book is read to the people. . . . Wisdom had all the city walls adorned with fine paintings representing all sciences. . . . Within the sixth circle are depicted the mechanical arts and their instruments, and how the various nations utilize them. Each art is ordered and explained according to its value, and also shows the name of the inventor.

This may be considered as a predecessor of today's scientific and technical data bases!

During the era of illuminism, which culminated in Diderot's *Encyclopedie* (1751–72), emphasis continued to shift from science to technology, which was applied to the industrial arts. According to Diderot, science pertains to the genius that creates, technology to sagacity that perfects. The utopian libertarian Prince Kropotkin (1842–1921) dreamt of a permanent education in the form of "research plus action" that would bring together men gifted with inventive ability (entrepreneurs?) but lacking scientific education and means for experimentation, with well-educated men with experimental facilities but lacking inventive genius (academicians?). Thus arose gradually the modern concept of a technopolis, which synergistically combines scientific research and invention with the practical applications of technology through the process of innovation.

Not all technopoleis, however, are perfect cities. The last example is Fritz Lang's film *Metropolis* (1926), which depicts life in a two-tiered city: the Aristopolis of skyscrapers and gardens dominated by a "control tower" that is the political and economic center of the city, and the underground city where the regimented workers live in barrack-like apartments and work at repetitive tasks in geometric factories. The same division of labor and differences in life styles are apparent, albeit to a much lesser degree, in Silicon Valley (Rogers and Larsen 1986: ch. 11).

It is important to note that all these utopian technopoleis are located in unknown or mythical countries, such as Atlantis, or in virgin lands with plenty of sun, pleasant surroundings and attractive climate. In fact, Morris's futurist industrial city is located in *Nowhere Land* (Morris 1891). Modern technopoleis have also risen in industrially virgin lands, or at least in undeveloped areas away from metropolitan centers: Route 128 was originally agricultural land, mostly dedicated to pig-farming; Silicon Valley was almond and plum orchards; Sophia-Antipolis, Mediterranean bush forest; the RPI Technology Park, pasture and woods on the banks of the Hudson River.

Comparing the mythical technopoleis of the past with the real technopoleis of modern times, we observe the same driving forces:

1. A visionary and determined leader,
2. A master plan for development,
3. Emphasis on scientific research and applications of science and technology,
4. Emphasis on technical education,
5. Rewards and honors for creativity, invention, and innovation,
6. Location in attractive areas with ideal environment,
7. Inducements and privileges to attract desirable members to the community and keep them,
8. Trained artisans and workers,
9. Political and social leadership by an elite, with the consensus of the entire community.

We will see later that the same driving forces have been instrumental in the creation of the developing technopolis of New York's Capital Region. This technopolis has its roots in

1. The rise and decline of the leading industry in the area, the General Electric Company,
2. The evolution of the role of the leading technical university, Rensselaer Polytechnic Institute.

THE RISE AND DECLINE OF INDUSTRY IN NEW YORK'S CAPITAL REGION

New York's capital region comprises the cities of Albany, Schenectady, and Troy and the neighboring communities. Its importance as a major communications hub emerged after the wars of independence, with the Erie Canal, the Champlain Canal, and the "water level" railroad route from New York City to the Great Lakes and the West. Parallel with the development of the communications infrastructure came the development of major industries. The valley of the Hudson River in the capital region area has relatively high and steep sides. Thus the many rivers and *kills* (a Dutch word for a fast flowing torrent in a side valley) were ideal sources of power to run the textile mills, machine shops, and iron works that flourished in the region. Cohoes, a few miles northwest of Troy, became a major textile center, Troy was named the "collar city," and the Watervliet Arsenal became a major supplier of ordnance during the Civil War. The American Locomotive Company was started in 1847 as the Schenectady Locomotive Works and produced more than 80,000 "iron horses" during its lifetime.

The transition from water to electricity as the major source of power for U.S. industry represented a technical discontinuity and thus a danger for the capital region area, whose industry was based on water power and mechanical technology. Fortunately, Edison wished to integrate vertically and not only supply electric lighting and power to New York City but also build generators and motors. He decided to move his Edison Machine Works away from New York City, and in 1886 he purchased for $45,000 two vacant buildings erected by the McQueen Locomotive Company in Schenectady. The enlightened citizens of that city wished to diversify from the mechanical to the electrical industry and contributed $7,500 toward this sum. This may be considered as an early example of the inducements that modern technopoleis use to attract desirable tenants.

This was the origin of the General Electric Company, the major industrial employer in the capital region. Edison did not move his laboratory from Menlo Park to Schenectady, but as General Electric grew and expanded its product scope, the need arose first for an engineering laboratory and later for a research laboratory. Steinmetz, the electrical genius who discovered the law of hysteresis and developed the symbolic method of A-C current calculations and the theory of electrical transients, became chief engineer in 1894 and later created a General Engineering and Consulting Laboratory, which was dedicated primarily to product development and application. In 1900 the first industrial research laboratory in the world was established in Schenectady. As stated in the 1902 General Electric *Annual Report*:

> Although our engineers have always been liberally supplied with every facility for the development of new and original designs, it has been deemed wise . . . to establish a laboratory devoted exclusively to original research. It is hoped by this means that many profitable fields may be discovered.

The key words are *original research, discovery, profitability,* and *many fields* (that is, interdisciplinary research). This interdisciplinary approach was reinforced by the fact that the newly appointed director, Willis Whitney, was an MIT *chemistry* professor, who insisted on retaining his MIT position for four years. The original staff consisted of only three professionals: Whitney, Steinmetz, and Dempster, a young assistant. In 1965 the General Engineering and Consulting Laboratory and the Research Laboratory were combined into the present General Electric R&D Center, the major research facility in the capital region, which presently employs about 2,200 people, of which almost 1,200 have technical degrees and 500 have doctorates. Famous engineers and scientists have gained international recongnition for the Center: Ernst Alexanderson, pioneer of wireless telephony; William Coolidge, inventor of the X-ray tube; Nobel Laureates Irving Langmuir, surface chemistry; and Ivar Giaevaer, superconductivity.

A second major research facility in the capital region is the Sterling-Winthrop Research Institute, devoted to pharmaceutical research and development. Founded in 1946 it employs about 900 people, of which about 450 are professional and 175 have Ph.D. degrees.

General Electric's employment level in the capital region reached the peak of 42,000 after World War II. It has been going down ever since and is presently at 13,000. In 1986 alone, 4,000 jobs were lost. The Gas Turbine Headquarters has moved to South Carolina, and the Large Motor and Generator Division has been shut down. Unfortunately, this decline of General Electric in Schenectady has coincided with an employment "bust" in other local major industries. A few examples will suffice:

1. The last textile factory in the Cohoes/Troy area closed down in 1982. Castle-like empty mills, often crumbling, remain as symbols of a bygone era.

2. Ford closed its Green Island radiator plant in 1983 on the basis of a technological discontinuity—namely, a change from copper to aluminum for radiators.

3. Sterling Corporation has moved some of its production facilities from the Greater Albany area to Puerto Rico, with an estimated loss of 300 jobs.

4. The local press has reported that employment at the Watervliet Arsenal will be cut from 2,600 to 1,100.

5. RPI is now the largest private employer in Troy, and the third largest in the capital region (after General Electric and a supermarket chain).

The reasons for this precipitous decline in industrial and related employment are varied and, in some cases, controversial: geographical decentralization, excessive union demands and strikes, the energy crisis, shifting world markets, foreign competition, obsolescent facilities, high taxes, and the transition from sunset to sunrise industries.

Parallelling the decline of industrial employment in the capital region has been a significant increase in employment by services, government, trade, and small businesses in general. Although these new jobs are welcome, their average salary level is 70 percent of manufacturing, their wealth creation is relatively low, and their export potential is minimal. Manufacturing companies remain the backbone of the U.S. economy and the guarantors of U.S. global competitiveness. Thus, with the transition of the U.S. to a postindustrial society, the capital region is evolving toward the two-tiered society, described in mythical technopoleis discussed above:

1.   An elite of Ph.D.s working in research institutes and the major universities in the area,
2.   Unskilled workers, employed in low value-added repetitive jobs and services, or unable to find stable employment.

In the meantime, the blue-collar skilled working class, which has always been the link between the two classes and the backbone of local citizenry, is gradually but steadily disappearing, an ominous sign for the economy of New York's capital region.

THE EVOLUTION OF RPI'S ROLE AS THE DRIVING FORCE FOR THE DEVELOPMENT OF A NEW TECHNOPOLIS

The concept and the realization of a technopolis depends on many driving forces. In today's terms, two of the most important forces are

1.   Innovative ideas that can be first synthesized in a vision by a leader or champion and then translated into practical implementation,
2.   A continuous flow of entrepreneurially creative, professionally competent and managerially adept people who will build up the technopolis according to the vision of the leader.

Traditionally, universities have been the source of both advanced ideas and creative, competent people. Rensselaer Polytechnic Institute (RPI), a private nonprofit technical university founded in 1824 and the oldest engineering school in the English-speaking world, has emerged as a "role model" (Abetti and

Le Maistre 1987) for proactively promoting industry/university cooperation, industrial innovation, and regional economic development, which in turn, have become the foundations of the emerging technopolis of New York's capital region. The evolving role of RPI is best viewed within three distinct time periods or "eras" (Abetti, LeMaistre, and Wallace 1986) corresponding to different phases in the economic development of the area.

Education and Knowledge Expansion (1824–1940)

From 1824 to 1940 the local economy was driven by infrastructure and industrial development, fueled by the growth of General Electric and its Research Laboratory, and RPI's main role was technical education and expansion of the knowledge base. RPI was one of the first engineering schools in the United States to set up an Electrical Engineering Department, and its graduates have reached high-level technical and managerial positions in large technology-intensive U.S. corporations, such as General Electric, IBM, AT&T, United Technologies, Kodak, and Colt Industries. Until World War II, the relationship of RPI with these major industries was relatively simple and straightforward: RPI furnished its best graduates to these and other companies, which, in turn, encouraged their officers to serve on RPI's board of trustees and assisted RPI with donations, equipment, fellowships, and occasional research grants. But—and this is a very important point—industrial innovation took place primarily *within* these companies, and there was limited *direct* technology transfer from RPI to industry.

R&D and Technology Transfer to Large Corporations (1941–80)

From 1941 to 1980 the economy was driven by a technology boom, which was followed by a bust in industrial employment. RPI added to its primary role of education that of applied research and development and technology transfer, mainly to large and medium-size established companies.

World War II created a new role for the leading U.S. technological universities. Technology was needed to win the war, to rebuild the infrastructure of destroyed nations, and to reconvert U.S. industry from defense to civilian uses. Naturally, major emphasis was given to applied research and development and industrial innovation, rather than basic "academic" research. Because of the unprecedented explosion of new science and technology, it became clear that even the largest U.S. high technology corporations could not be expert in all the scientific disciplines and core technologies that were (or, more important, could become) the driving forces for their rapidly expanding product lines, manufacturing processes, and new services. Thus, progressive companies naturally turned to the technological universities as sources of technology, to be developed through cooperative university/industry initiatives. The National Science Foundation encouraged such initiatives and funded the establishment of centers to develop leading-edge technologies at selected universities. RPI's Center for Interactive Computer Graphics was started in 1977 with a NSF grant of $400,000. It is now considered as the leading world center in this field, with fifty researchers, thirty-eight sponsoring or member companies, and a budget of $2 million per year. The center focuses its R&D on computer-aided design, solid geometric modeling, finite element analysis, and in general problems associated with the application of computers in design and management.

About the same time, it was recognized that the U.S. manufacturing industry was losing its worldwide competitiveness, an indication being that productivity gains in the United States were much lower than in Europe and Japan. Because of the conservatism and inbreeding of many U.S. mechanistic and segmented manufacturing organizations, the *transfer* of new technologies—such as robotics, computer-assisted manufacturing and testing, and flexible automation—was particularly difficult. The RPI Center for Manufacturing Productivity and Technology transfer was created in 1979 to

1. Develop innovative methods to enhance U.S. manufacturing productivity and regain U.S. international competitiveness,
2. Transfer advanced manufacturing technology to member companies,
3. Revive the interest of faculty and students in the manufacturing technologies and in the strategic role of manufacturing within the firm.

This center is performing approximately $3 million of research and development in financial year 1986. Projects cover the range of manufacturing with a major concentration in robotics and computer-integrated manufacturing (CIM).

About 1980 integrated electronics emerged as perhaps the most important technology of this century. The major boom in this glamorous field has taken place in the technopolis of Silicon Valley, around Stanford University. RPI realized that eastern-based companies, such as AT&T, General Electric, IBM, and Digital Equipment, could benefit from a multidisciplinary center for advancing the state of the art and created in 1981 the Center for Integrated Electronics. This center is truly interdisciplinary, involving faculty from the School of Engineering (Materials and Electrical, Computer, Systems Engineering Departments) and the School of Science (Physics and Chemistry Departments). Support has been provided by industry and the Semi-Conductor Research Council. Areas of research encompass both silicon and gallium arsenide applications. Specific areas include the use of expert systems applied to the computer aided design of electronic circuits and a major specialization in the application of beam technology.

These three major R&D centers are making a major contribution to industrial innovation, thanks to their close coupling with their industrial sponsors. It is well known that technology is primarily transferred by *people* (rather than paper), and at RPI this is accomplished by

1. Frequent meetings with the sponsoring companies, which participate in setting the goals of R&D to be carried out by the three centers,
2. Company engineers in residence for shorter and longer terms, up to several years, at the centers,
3. Offers of part-time and full-time employment in the companies to researchers and students working in the centers.

In order to consolidate the roles of the three centers as driving forces for industrial innovation and to integrate their activities with industrial engineering and technology management, the late president of RPI, George Low, proposed the creation of a center for Industrial Innovation, which would act as a magnet and role model for the entire state of New York. The response of the leading industrial firms based in New York state, such as GE, IBM, Kodak, and Colt Industries was enthusiastic, and they supported RPI's proposal to New York state. The result was that the state legislature granted RPI a $30 million interest-free loan to construct a building on campus to be called the Center for Industrial Innovation (CII)—the loan to be repaid over a forty-year period. The three centers described earlier are being relocated into this building, of 200,000 sq. ft., including laboratories, offices, lecture theaters, and 8,000 sq. ft. of clean rooms. RPI has a commitment to equip this building with $30 million worth of hardware and much of this is now installed. In addition, the CII, on the basis of the enabling legislation, must develop an outreach program to New York industry. Interaction with faculty from community colleges has occurred (sabbaticals), and an off-campus master's degree is under development. Courses will be delivered to these sites via satellite. Thus, the CII and its three centers became the first seed of the technopolis.

Technological Entrepreneurship and Regional Economic Development (1981–present)

From 1981 to the present, while the industrial employment bust has continued unabated, the main economic driving forces have shifted to services, small businesses, and new high tech ventures. Consequently, RPI has added to its roles of education, research, and technology transfer, the new proactive role of promoter of technological entrepreneurship and regional economic development through job creation by new high-value-added companies and thus has become the driving force for the development of the new technopolis.

This loss of local jobs, coinciding with the technological boom described earlier, is not altogether surprising. As we discussed, technology transfer from RPI and other R&D institutions took place primarily to large and medium-size firms because only these corporations maintained traditional corporate relationships with academia, funded research, and could afford membership in the various technology centers. It is these same large corporations that have decreased their total employment in the United States since 1975, and in the case of the capital region there is no hope of reversing the precipitous decline of industrial employment by appealing to the large local mature firms to remain in the area or create new jobs.

In contrast, new entrepreneurial high tech ventures are inherently high-value-added firms and major creators of new jobs (Morse and Flender 1976), either by direct employment or by "pull-through" of supporting jobs in supply and service firms. Figures ranging from five to ten low tech jobs for every high tech job have been quoted. Also, small firms are considered to be more innovative and more efficient in the use of scarce human and material resources than large firms, including bringing innovations faster to market. These firms are usually created by entrepreneurs who, if the climate is favorable, prefer to remain in the geographic area where they started their business (Vesper 1983).

George Low was firmly convinced that RPI should assume an active rather than passive role in the community and should take leadership in attracting and assisting entrepreneurs who wanted to create and develop new high tech companies. With this vision, he himself acted as an entrepreneur in committing RPI, a private institution, to invest several million dollars of its funds to launch the Incubator Program and the Rensselaer Technology Park, which became the second and third seeds of the new technopolis.

THE RPI INCUBATOR PROGRAM

From its austere beginnings in 1980 as an experiment located in the basement of a classroom building, the RPI Incubator Program has become a nationally recognized model for university-supported business development. In less than five years, the program has spawned dozens of new enterprises. Many have raised large sums of capital (including venture capital, private placements, and public offerings), and all have had a positive impact on the local economy and RPI's academic environment.

The Incubator Center provides the support and resources that are critical for new business ventures:

1. The rent is based on a graduated payment structure that allows younger companies to pay less.
2. A receptionist, answering service, and photocopying equipment are available on the premises for any company to use.
3. Companies have direct access to RPI's computers, libraries, data base systems, shops, laboratories, testing equipment, faculty consultants and student assistance in the following areas:

CAD/CAM	Microbiology
Integrated electronics	Biomedical engineering
Interactive computer graphics	Chemical process design
Automation and manufacturing	Tribology and wear
Robotics	Control theory
Composite materials	Management, marketing, and
Quality assurance	legal assistance

Any individual or existing company may join the Incubator Program if it meets the four criteria for admission:

1. It intends to develop an innovative technical idea;
2. It intends to serve an existing or potential market;
3. It has or intends to develop effective linkages with RPI faculty and students;
4. It is able to pay rent and willing to let RPI take 2 percent equity participation.

The four criteria are designed to select for the Incubator Program companies that have a high probability of *business* (in contrast to technical) success.

1. The first criterion rejects nontechnical companies that do not belong in a technical university and "me too" companies that are not innovative and therefore have a lower probability of success.

2. The second criterion rejects those would-be entrepreneurs who are driven by technology alone (the "better mousetrap" syndrome) and have given no thought to the needs of the marketplace.
3. The third criterion rejects companies that are looking only for low-cost rental space and access to RPI resources. It assures that there will be synergy between the company and RPI students (part-time, summer, and full-time employment) and faculty (technical and management consulting, research on entrepreneurship, and so forth). In fact, the major benefit to RPI from this program has been this close coupling between academia and business, where students have the opportunity of working under the direction of a company president or vice-president of engineering.
4. The fourth criterion ensures that RPI, a private, nonprofit institution, will recover its costs (except depreciation on the building) for the services rendered. Initially, no equity was required from participating companies. Should one of the incubator companies become another Xerox, Polaroid, DEC, Data General, or Apple, the symbolic 2 percent participation will contribute toward RPI's financial well-being.

RPI does not, as a rule, provide funds to the incubator companies. However, the director of the program, Jerome Mahone, is a former banker and has close connections with local banks and venture capitalists in the northeast. He helps the incubator companies write effective business plans and introduces them to sources of funds and capital. All local resources are listed in a recently published manual, prepared by the RPI School of Management for the Capital Region Technology Development Council (1987).

The results of the thirty-six incubator companies that formally joined the program since 1982 have been impressive. As of October 1986 fifteen have graduated successfully and moved to the Rensselaer Technology Park or to other facilities in the area (except for Raster Technologies, which moved to Route 128 in Massachusetts), fifteen are incubating, and four withdrew for various reasons (death of a principal, lack of entrepreneurial drive, lack of time or funds). So far, only two companies have failed.

The successful incubator companies represent varied technologies, including bioreactors, computer displays, robotics, software, magnetic resonance imaging systems, and nationwide game and information electronic networks. New companies are admitted at the average rate of one per month as space becomes available. Total employment created to date is estimated at 400 full-time and part-time jobs, of which 250 are in the area and 150 on Route 128. Aggregate sales of the fifteen graduated companies amount to approximately $25 million. The fifteen incubating companies compose the second generation. They employ forty-seven full-time personnel and thirty-eight students and interact with twenty-seven faculty members. Therefore, an employment ratio of approximately five to one is becoming evident relative to the number of jobs created by companies after graduation, as compared to new jobs by companies in the Incubator Program.

Whatever the yardstick, be it new jobs or the success rate of new ventures, or the transfer of technological ideas into products and services, or the "value added" to the educational environment of the university, the RPI Incubator Program has proven to be a great success and the accepted model for other programs.

RENSSELAER TECHNOLOGY PARK

Feasibility studies of the Rensselaer Technology Park (RTP) were initiated in 1979 by Michael Wacholder (now director of RTP), to convert several hundred acres of pasture and woodland owned by RPI in North Greenbush, about five miles south of the campus (and only fifteen minutes drive from Troy, Albany, and Albany Airport) into an environment for technology ventures. While these studies were in progress, the Incubator Program experienced significant success and it became apparent that the development of the Technology Park should proceed. In 1981 RPI decided to invest several million dollars of its endowment funds to develop the first 100 acres of the park, and the infrastructure work was completed in 1982.

To maintain strict control of the environment, Rensselaer Technology Park does not sell land but leases it on a long-term basis (forty-nine to ninety-nine years). All buildings and site development must conform to rigid specifications, and only R&D, service, and light manufacturing activities are allowed. The first tenant for the park was National Semiconductor's Optoelectronics Division in 1983. It had acquired a local company specializing in epitaxial GaAs light sources. National Semiconductor had to decide whether to move this company from Latham, New York, to California or whether to move its own plant from Silicon Valley to the Rensselaer Technology Park. When it found that housing and related costs in the capital region were one-third of those in Silicon Valley, the answer was clear.

For the graduating incubator companies and other small or medium-size companies, which did not require an entire building, RPI has built five multitenant buildings that offer short-term lease space. An additional incentive was developed, a creative funding program instituted by the town of North Greenbush, where HUD Small Cities Block Grant funds are utilized for seed capital investments (up to $100,000) in young companies that plan to increase employment.

The results of the first four years (1983 through 1986) of the Rensselaer Technology Park have been very rewarding. Thirty-four companies are located in the park, including six graduates from the Incubator Program, and together they employ about 350 people. Employment will increase to 700 in the spring of 1987, when NYNEX (New York and New England Telephone Company) will occupy its new data center in the park. About 90 percent of the companies have already demonstrated interactions with RPI faculty, students, and resources.

Clearly, there is a trend for technological universities in the United States, Europe, and the Orient to develop closer ties with the corporate community that they are serving. The technology or research park is becoming a popular and fast-growing manifestation of this trend, and RPI has become well recognized as a second-generation pioneer (following the earlier successes of Stanford/Silicon Valley and the Research Triangle in North Carolina). Unlike many other parks that are being developed around a singular technology (that is, biotechnology or electronics), the RPI model is focused on diversity and strong ties between park companies and the vast array of resources at the university. The range of technologies already in the park include electronics, robotics, composites, telecommunications, biotechnology, adhesives, and software.

UNIQUE CHARACTERISTICS OF THE DEVELOPING TECHNOPOLIS

The developing technopolis of New York's capital region differs from other U.S. technopoleis in two important characteristics.

Timing

Research and industrial parks created by or associated with major universities may be considered as the "anchors" around which technopoleis are growing or expected to grow. It is interesting to look at the number of parks that created each quinquennium, from 1951, when Stanford Research Park was created, to 1975 (Cartier and de Kerorguen 1985: 42–54):

1951–55	5
1955–60	18
1961–65	46
1966–70	9
1971–75	3

It appears that the rate of growth of science parks, just as other industrial investments, coincided with the growth phase of the last Kondratieff economic cycle, which peaked in 1968. On the contrary, Rensselaer Technology Park (RTP) was created and is growing in a period of stagflation, which corresponds to the trough of the Kondratieff wave, expected to occur around 1989. According to Mensch (1979: ch. 7) the decade 1984 through 1994 will show a major surge of technological innovation, and thus we may conclude that Rensselaer Technology Park (RTP) was established at the optimum time.

Location

Compared to the cities of the sun, New York state is not a particularly attractive location, as far as climate is concerned. Also, New York has one of the highest income tax rates of the fifty states, although this is gradually being reduced. We have also pointed out that the capital region is a mature area that is being abandoned by large manufacturing companies, with a precipitous decline in industrial employment. In contrast, many successful technopoleis have been created in regions with rapid economic growth such as the San Francisco Bay area, the Boston metropolitan area, the Austin/San Antonio area in Texas, the French Riviera, and so forth. Psychologically, it is much harder to create something in a climate of decline and decay than in the euphoria of growth and success.

It is well known that not all science and technology parks have been successful. Some failed to develop despite substantial private and especially public investments, others have grown very slowly and thus have not become the hoped-for foundations of new technopoleis. Even the now highly successful Research Triangle Park in North Carolina did not really "take off" until fifteen years after it was launched. So the question arises: What are the factors that have determined the success of the developing technopolis in New York's capital region? To answer this question we need to review the driving forces for the creation of technopoleis discussed at the beginning of this study.

CONCLUSION: DRIVING FORCES FOR
TECHNOPOLIS CREATION AND DEVELOPMENT

We have seen that there is continuity between the myth of the cities of light of the Renaissance and illuminism, and the modern technopoleis. We will now compare the driving forces of past myth and modern reality (Cox 1985)

Table 9–1. Analysis of Driving Forces for Technopolis Creation and Development.

Myth (1600–1800)	Reality (1950–87)	Application to RPI and New York Capital District
1. Visionary and determined leader	Executive champion	George Low, president of RPI
2. Master plan for development	Spontaneous growth (Route 128) versus development plan (North Carolina Research Triangle)	Strategic Plans for RPI: Center for Industrial Innovation Incubator Program Technology Park RPI 2000
3. Emphasis on scientific research and applications of science and technology	Major institutional research facilities	General Electric R&D Center, Sterling-Winthrop Research Laboratory, etc.
4. Emphasis on technical education	Major technical university, heavily engaged in applied research	Rensselaer Polytechnic Institute, State University of New York at Albany, Albany Medical Center, etc.
5. Rewards and honors for creativity, invention, and innovation	Peer recognition, awards, patents Venture financing Entrepreneurial opportunities and rewards	Nobel prize, awards, grants Use of RPI patents Funding from local and state government Assistance to entrepreneurs
6. Location in attractive areas with ideal environment	Warmer climates preferred, desirable living environment	Colder climate High taxes Pleasant living environment

7. Inducements and privileges to attract desirable members to the community	A great variety of local and state government incentives, local venture capital pools, incubators, etc.	Initially very few incentives (except low rent in Incubator) Now varied sources
8. Trained artisans and workers	Skilled labor force, technicians	Many skilled workers laid off by the major companies in the area / Graduates of two-year technical institutes
9. Political and social leadership by an elite, with the consensus of the entire community	Initiatives of academic, technical business and political leaders, with support from all local constituencies	Support by New York state governor and political leaders / Capital District High-Tech Council / General community consensus

and determine how these forces have contributed to the creation and development of the technopolis of New York's capital region, as spearheaded by RPI. Our analysis is summarized in Table 9–1. Two general conclusions may be drawn from this table:

1. Essentially the same driving forces of past myths drive the creation and development of modern technopoleis. This is not surprising because our present Western culture is the heritage of humanism, renaissance, and illuminism.
2. Most of these forces, but not necessarily all, need to be present for the successful creation and development of a modern technopolis. For instance, a warm climate is not a prerequisite, if offset by other forces.

In the specific case of New York's capital region, the key driving forces have been

1. An executive champion, George Low, the late president of RPI,
2. The proactive strategic role of RPI in developing the Center for Industrial Innovation, the Incubator Program, and the Technology Park,
3. The presence of major research institutions in the area (such as General Electric and Sterling-Winthrop),
4. A continuous stream of creative technical people and entrepreneurs from the universities and industries in the area, heavily engaged in applied research,
5. The availability of skilled workers, laid off by major companies in the region,
6. A tradition for innovation, established by the technical, academic, business, and political leaders, and supported by the consensus of the entire community.

A recent study of the key factors influencing the location decision process of entrepreneurial companies at Rensselaer Technology Park (RTP) appears to substantiate the importance of the above factors.

Entrepreneurial (and intrapreneurial) companies, particularly if considered high tech, are eagerly sought after as desirable tenants by a variety of private and public real estate organizations and economic or industrial development authorities. Therefore, the principals of such companies face a bewildering spectrum of choices in selecting prospective permanent locations. From discussions with entrepreneurs or intrapreneurial managers of newly created branches of established companies, a fifty-item questionnaire was developed at

RPI to determine the key factors that influence the location decision process of U.S. and Irish entrepreneurs (Abetti et al. 1987).

The questions were grouped into five categories:

1. *Company demographics* (employees, revenues, space, years in operation, type of company),
2. *Financial considerations* (lease rates and terms, space availability, cost of employees, and so forth),
3. *Location* (geographic location; buildings and grounds; area image; transportation; employee skills; interaction with customers, suppliers, services, government; recreational and cultural activities; and so forth),
4. *Funding* (local, state, federal government funding; tax incentives; private sources, banks, and venture capital),
5. *University connections* (technical and management support, personal ties, students as employees, faculty consultants, image).

The data were collected in July 1986 from twenty-three tenants of Rensselaer Technology Park (out of twenty-eight total) and led to some significant conclusions (in parenthesis are the percentages and standard deviations of respondents considering the category factors as "very important" or "important"):

1. Even though the purpose of the Rensselaer Technology Park is different from that of a conventional real estate project, the success of the park is based on the recognition that the real estate considerations must be competitive. Therefore, financial factors have the greatest influence in the entrepreneurs' location decision process. (72 ± 17 percent)
2. The RPI connection is next in importance. This category differentiates RTP from all other developments in the area. (57 ± 13 percent)
3. Location is considered less important, despite the fact that this factor is rated as key by commercial real estate developers. (48 ± 17 percent)
4. Location to local funding sources is the least important. This apparently surprising result may be explained by the fact that there are limited funding sources for new ventures in the New York capital region, while venture capital is available from major nearby financial centers, specifically New York City and Boston. (35 ± 9 percent)

In conclusion, the creation and development of a technopolis represents a major social and technical innovation (Abetti 1986), which will have profound effects on the economic, social, and cultural life of the community. We believe that the RPI initiative, described in this paper, provides a model for such technopolis development, which could be followed by other mature industrial areas in the world, to regain their industrial strength and economic well-being.

BIBLIOGRAPHY

Abetti, P.A. 1986. "Innovation from Start to Finish." *Chemtech* 16 (July): 405–12.

Abetti, P.A., and C.W. Le Maistre. 1987. "Rensselaer Polytechnic Institute as a 'Role Model' for Promoting Industrial Innovation, Manufacturing Productivity and Regional Economic Development." In *Industrial Innovation Productivity and Employment*, edited by P.A. Abetti, C.W. Le Maistre, and R.W. Smilor, pp. 63–79. Austin: University of Texas at Austin, IC^2 Institute Monograph.

Abetti, P.A., C.W. Le Maistre, and W.A. Wallace. 1986. "The Role of Technological Universities in Nurturing Innovation: The RPI 'Model.' " In *Technological Innovation: Strategies for a New Partnership*, edited by D.O. Gray, T. Solomon, and W. Hetzner, pp. 251–60. Amsterdam: Elsevier (North-Holland).

Abetti, P.A., J. O'Connor, L.M. Ehid, J.L. Rocco, and B.J. Sanders. 1987. "A Tale of Two Parks." Paper presented at the Seventh Annual Babson College Entrepreneurship Research Conference, Wellesley, Mass., April 29–May 1.

Bacon, Francis. [1627] 1942. "*New Atlantis.*" In *Essays and New Atlantis*, pp. 243–302. Roslyn, New York: Walter J. Black and Company.

Campanella, T. 1623. *Civitas Solis*. Frankfurt: Typis E. Emmelii. In Latin.

Capital Region Technology Development Council. 1987. *1987 Finance Resource Manual*. Albany, N.Y.: CRTDC.

Cartier, A., and Y. de Kerorguen. 1985. "Technopoleis, Status Report." In special issue of *Technopolis*, pp. 42–54. Paris: Autrement Revue.

Cox, R.N. 1985. "Lessons from 30 Years of Science Parks in the USA." In *Science Parks and Innovation Centres: Their Economic and Social Impact*, edited by J.M. Gibb, pp. 17–24. Amsterdam: Elsevier.

de Kerorguen, Y. 1985. "Science Parks and Radiant Cities." In special issue of *Technopolis*, pp. 234–38. Paris: Autrement Revue. In French.

Mensch, G. 1979. *Stalemate in Technology*. Cambridge, Mass.: Ballinger.

Morris, William. [1891] 1914. *News from Nowhere*. New York: Longman's Green and Company.

Morse, Richard, and John O. Flender. 1976. *The Role of New Technical Enterprises in the U.S. Economy*. Washington, D.C.: U.S. Government Printing Office, Technical Advisory Board, U.S. Department of Commerce.

Rogers, E.M., and J.K. Larsen. 1986. *Silicon Valley Fever*. New York: Basic Books. Esp. Chapter 11, "Problems in Paradise," pp. 184–202.

Vesper, Karl H. 1983. *Entrepreneurship and National Policy*. Chicago: Heller Institute for Small Business Policy, Paper 3.

Chapter 10

THE AUSTIN/SAN ANTONIO CORRIDOR
The Dynamics of a Developing Technopolis

Raymond W. Smilor, George Kozmetsky,
and David V. Gibson

This chapter examines a number of critical components in the development of the Austin/San Antonio Corridor as a technopolis. The research traces the most important events in a number of sectors from 1945 through the end of 1986. During this period of time, the Corridor moved from being an essentially small university town within a state capital on the north and a military dependent town on the south to a developing technopolis within a region of relative general economic slowdown. This research seeks to describe the most important environmental forces, organizational issues, key individuals, and public/private sector relationships that contributed to the growth and downturn of this technopolis.

This chapter also develops a conceptual framework, which we call the *Technopolis Wheel,* to describe the process of technology development and economic growth in the Austin/San Antonio Corridor. We believe that this concept of the Technopolis Wheel has important implications for the development of other technopoleis in the United States and perhaps, to some degree, other nations. In the United States the Wheel reflects the interaction of seven major segments in the institutional make-up of a technopolis. These seven segments include the university, large technology companies, small technology companies, state government, local government, federal government, and support groups. Finally, and perhaps most important, this chapter considers key individuals, or influencers, who link the seven segments of the Wheel (see Figure 10–1).

Figure 10-1. The Technopolis Wheel.

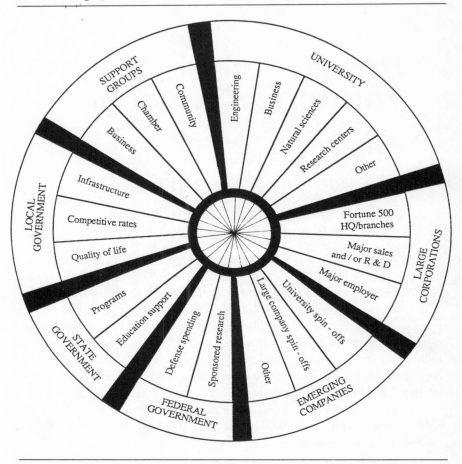

New kinds of institutional developments between business, government, and academia are beginning to promote economic development and technology diversification. Several studies have begun to analyze these new relationships and organizational structures (see Allen 1986; Harrigan 1985; Hisrich 1986; Konecci 1985; Krzyzowski 1982; Ouchi 1984; Rogers 1984; Ryans 1986; Watkins 1985). They have sought to evaluate how business, government, and universities interrelate, particularly concerning technology development and diversification.

A fascinating paradox has emerged—the paradox of competition and co-operation in the development of a technopolis (Ouchi 1984). On the one hand, a great deal of competition takes place between universities, companies, and public- and private-sector entities. On the other hand, cooperation is essential for a technopolis to develop. Segments within the technopolis must find new ways to cooperate while competing. This study looks at some of the dynamics of this paradox of competition and cooperation and the new kinds of institutional developments that have emerged to deal with it. The conceptual framework of the Technopolis Wheel seeks to explain some of the components of these new relationships for competition and cooperation. It focuses on the concept of networking—that is, the ability to link public- and private-sector entities, some of which have been traditionally adversarial, to effect change.

DEFINING AND MEASURING THE TECHNOPOLIS

In the United States the modern technopolis is one that interactively links technology development with the public and private sectors to spur economic development and promote technology diversification. Three factors are especially important in the development of a technopolis and provide a way to measure the dynamics of a modern technology city-state:

1. *The achievement of scientific preeminence.* A technopolis must earn national and international recognition for the quality of its scientific capabilities and technological prowess. This may be determined by a variety of factors including R&D contracts and grants; membership of faculty and researchers in eminent organizations such as the National Academy of Sciences and the National Academy of Engineering; the number of Nobel Laureates; and the quality of students. In addition, scientific and technological preeminence may be measured through new institutional relationships such as industrial R&D consortia and research and engineering centers of excellence.
2. *The development and maintenance of new technologies for emerging industries.* A technopolis must promote the development of new industries based on advancing cutting-edge technology. These industries provide the basis for competitive companies in a global economy and the foundation for economic growth. They may be in the areas of biotechnology, artificial intelligence, new materials, and advanced information and communication technologies. This factor may be measured through the commercialization of university intellectual property, and new types of academic/business/government collaboration.

3. *The attraction of major technology companies and the creation of home-grown technology companies.* A technopolis must affect economic development and technological diversification. This may be determined by the range and type of major technology-based companies attracted to the area, by the ability of the area to encourage and promote the development of home-grown technology-based companies, and by the creation of jobs related to technologically based enterprises.

BACKGROUND OF THE CORRIDOR

The Austin/San Antonio Corridor is a strip of land approximately 100 miles long and twenty-five miles wide in the heart of Texas (Figure 10–2). Interstate Highway 35 connects Austin on the north end of the Corridor to San Antonio on the south end. To the east side of the highway lies the Blackland Prairies, some of the richest farmland in the United States. To the west of the highway lies the famous Hill Country of Central Texas.

The seeds of the Austin/San Antonio Corridor's development go back at least to World War II. During the early and mid-1940s the U.S. Air Force and Army expanded their presence in both San Antonio and Austin. That presence contributed to the early stages of economic growth in the Corridor.

In 1950 a string of seven dams was completed and created the Highland Lakes in Central Texas. The resulting lakes provided a major improvement in the quality of life, particularly in recreation, and helped reshape the geography and perception of the area. Throughout the 1950s and early 1960s infrastructure improvements, such as the development of new airports and highways, were also undertaken. When Lyndon Baines Johnson became president in 1963, increasing national and international attention was focused on the area of Texas where he was raised: the Hill Country of Central Texas.

In the 1970s rising oil prices benefited all of Texas including the Corridor and allowed an expansion of state appropriations for higher education, which in turn advanced the development of the technopolis. By 1980 the rate of population growth in the Corridor was 2.5 times that of the United States. By the mid-1970s the increasing growth in Austin and San Antonio began to raise questions about growth management and the benefits of growth in general.

The early 1980s were special years for Texans because of the state's approaching sesquicentennial in 1986 and centennial celebrations at the state's two flagship universities: The University of Texas at Austin and Texas A&M University. The development of the Austin/San Antonio technopolis reached a

Figure 10-2.   The Austin/San Antonio Corridor.

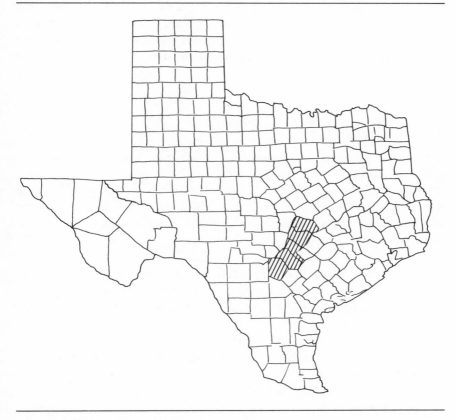

Source: The Austin/San Antonio Corridor Council.

crescendo in 1983 when MCC chose Austin as its headquarters after a major and very public site selection process among some of the most visible high tech centers in the United States. Austin made headlines in the *New York Times,* the *Wall Street Journal,* and the world press as the next great Silicon Valley. Nicknamed Silicon Prairie, Silicon Gulch, and Silicon Hills, the area generated an unprecedented wave of enthusiasm because of the perception that it had suddenly become a major technology center.

In 1984 the dramatic and unexpected plunge in oil prices coupled with declining farm and beef prices caused a general economic decline: a state that previously enjoyed a budget surplus and no corporate or personal income taxes now faced budget deficits. The Austin/San Antonio Corridor in terms of the development of the technopolis began to loose its momentum. Between 1984 and 1987 the Corridor began to experience a series of problems revolving around a general economic recession in the state, cutbacks in higher education funding, changes in local governmental attitudes, and a speculative development cycle that ended in foreclosures and bankruptcies and a general loss of direction.

THE UNIVERSITY SEGMENT

The nucleus in the development of the Austin/San Antonio Corridor as a technopolis has been the university segment. In Austin, at the north end of the Corridor, The University of Texas at Austin has played the key role. In San Antonio, at the south end, The University of Texas Health Science Center and The University of Texas at San Antonio have played key roles.[1] These universities have been pivotal in several ways:

1. By fostering research and development activities;
2. By contributing to perceptions of the region as a technopolis;
3. By attracting key scholars and talented graduate students;
4. By fostering the spinoffs of new companies;
5. By attracting major technology-based firms;
6. By nurturing a large talent pool of students and faculty from a variety of disciplines;
7. By acting as a magnet for federal and private-sector funding; and
8. By providing a source of ideas, employees, and consultants for high technology as well as infrastructure companies, large and small, in the area.

Indeed, the fundamental point can be made that if these major research universities were not in place and had not attained an acceptable level of overall excellence, then the Corridor could not have begun to be developed as a technopolis. There would be little or no research and development funding, no magnet for the attraction and retention of large technology-based companies, and no base for the development of small technology companies.

The University of Texas at Austin now claims two Nobel Laureates in physics and thirty-eight faculty members who belong to the National Academies of Science and Engineering. In addition, the number of National Merit Scholar students has continued to rise from 361 in 1981 to 916 in 1986. In 1986 UT was second in total national scholarship graduates with only eighty-seven fewer than Harvard.

The total dollar amount of contracts and grants (both federal and nonfederal) awarded to UT has increased steadily by year from 1977 (about $55 million) to 1986 (about $120 million). Much of the increase can be attributed to the UT Endowed Centennial program for chairs, professorships, and fellowships. In other words, centennial endowments have made a significant difference in attracting researchers who in turn attract additional research funds.

Nonfederal funding to the university has also increased. Nonfederal includes industrial, foundation, and state sources of support. The number of nonfederal contracts and grants from 1977 to 1986 has grown from 181 to 485. The dollar amount of these contracts and grants has grown from over $7 million in 1977 to nearly $25 million in 1986.

The university has established major organized research units in the College of Engineering and College of Natural Sciences. Table 10–1 shows eighteen major research centers in the College of Engineering with a total funding in 1986 of $28,916,099. Table 10–2 shows thirty-two research centers in the College of Natural Sciences with a total funding in 1986 of $21,354,719. Many of these research units are in emerging, cutting-edge technological areas.

Table 10–1. Organized Research Units in Engineering at UT/Austin, January 1987.

	Funding Levels
Aeronautical Research Center	$ 415,491
Texas Institute for Computational Mechanics	415,018
Computer and Vision Research Center	399,487
Construction Industry Institute	1,425,519
Center for Earth Sciences and Engineering	253,266
Electrical Engineering Research Lab	941,150
Center for Electromechanics	11,096,384
Electronics Research Center	699,255
Center for Fusion Engineering	460,081
Geotechnical Engineering Center	461,975
Center for Materials Science and Engineering	1,709,255
Microelectronics Research Center	2,041,240
Center for Enhanced Oil and Gas Recovery Research	691,355
Center for Polymer Research	1,356,817
Center for Space Research and Applications	1,558,102
Phil M. Ferguson Structural Engineering Laboratory	859,081
Center for Transportation Research	3,224,539
Center for Research in Water Resources	908,084
Total: 18	$28,916,099

Table 10-2. Organized Research Units in Natural Sciences at UT/Austin, January 1987.

	Funding Levels
Center for Applied Microbiology	$ 751,158
Artificial Intelligence Laboratory	1,422,556
Institute for Biomedical Research	515,489
Brackenridge Field Laboratory, the Field Station	204,422
Cell Research Institute	422,686
Central Hybridoma Facility	
Clayton Foundation Biochemical Institute	1,154,731
Institute for Computing Science and Computer Application	1,637,166
Culture Collection of Algae	153,005
Center for Developmental Biology	520,446
Laboratory of Electrochemistry	551,852
Center for Fast Kinetics Research	727,458
Fusion Research Center	5,507,251
Institute for Fusion Studies	2,500,287
Genetics Institute	1,293,788
Ilya Prigogine Center for Studies in Statistical Mechanics	400,311
Center for Materials Chemistry	
Center for Nonlinear Dynamics	424,378
Center for Numerical Analysis	186,719
Center for Particle Theory	225,094
Plant Resources Center	53,681
Protein Sequencing Facility	
Radiocarbon Laboratory	86,896
Institute for Reproductive Biology	520,323
Center for Relativity	279,081
Research Institutes-Weinberg	
Research Instruments Laboratory	75,277
Center for Statistical Sciences	149,281
Center for Structural Studies	300,297
Institute for Theoretical Chemistry	863,617
Theoretical Physics	252,603
Vertebrate Paleontology Laboratory	174,866
Total: 32	$21,354,719

The number of endowed fellowships, lectureships, professorships, and chairs in The University of Texas at Austin has also increased significantly since 1981. Fellowships and lectureships in business have increased from two to sixty-eight. Fellowships and lectureships in engineering have increased from zero to sixty-seven. Fellowships and lectureships in natural sciences have increased from two to fifty. Professorships in business have increased from twenty to seventy-one. Thirty-two of the seventy-one professorships were filled by the end of 1986. Professorships in engineering have increased from thirty to fifty-nine. Fifty-one of the fifty-nine professorships were filled by the end of 1986. Professorships in natural sciences increased from twelve to seventy-five. Forty-three of the seventy-five professorships were filled by the end of 1986.

Chairs in business increased from three to twenty. Twelve of the twenty chairs were filled by the end of 1986. Chairs in engineering increased from seven to thirty-four. Twenty of the thirty-four chairs were filled by the end of 1986. Chairs in natural sciences increased from three to thirty-six. Thirteen of the thirty-six were filled by the end of 1986.

The University of Texas Health Science Center (HSC) at San Antonio is a health professions university and a leading biomedical education and research institute in the Austin/San Antonio Corridor. The university has 700 full-time faculty members, 3,400 employees, and 2,200 students. The university offers degrees in five schools: medical, dental, nursing, allied health sciences, and the graduate school of biomedical sciences. A cooperative Ph.D. degree is offered with The University of Texas at Austin's College of Pharmacy.

Since 1975 the Health Science Center has more than quadrupled its grants of research funds. It had more than $40 million in 1986 in sponsored research projects. These major research areas include cancer, cardiovascular disease, pulmonary and kidney disease, immunology, reproductive biology, aging, genetics, arthritis, nutrition, and psychiatry. The HSC has three centers that are nationally funded. These are the Multipurpose Arthritis Center, the Center for Research and Training in Reproductive Biology, and the Center for Development Genetics. In addition it has received a five-year grant from NSF to develop an Industry-University Cooperative Center for Bioscience and Technology.

In Texas the state government is responsible for the major portion of funding for the budgets of public universities. The University of Texas component institutions have also benefited tremendously from a Permanent University Fund (PUF), with a current book value at $2.6 billion. The fund has been crucial to the development of the teaching and research excellence at UT (and Texas A&M), as well as in permitting the acquisition of modern facilities and laboratories. The PUF alone, however, is insufficient for the development of a world-class university.

For example, in 1984, while oil prices were still about $30 a barrel and state revenues increased by $5.4 billion or 17 percent over the previous year, Texas was the only state in the nation to decrease appropriations for higher education, a decrease of 3 percent. In that same year, California increased its state appropriation for higher education by 31 percent over the previous year. It was at this point that UT's momentum toward teaching and research excellence, such as being able to fill endowed positions, began to slow down.

Consequently, despite UT's recent phenomenal growth in endowed chairs, professorships, lectureships, and fellowships; despite the location of MCC in Austin; and despite national and international press claiming Austin/San Antonio a new center of excellence in education, the lack of sustained state support for higher education sent a mixed message to the best scholars and researchers whom the university was trying to attract.

In summary, as state allocations for higher education increased through the late 1970s and the early 1980s, the perception of the development of the Austin/San Antonio Corridor as a technopolis outside the state increased proportionately as well. On the other hand, as the state of Texas began to cut back on its funding to higher education in 1983, the perception of the Corridor as a developing technopolis declined, and the perception of retrenchment in the university began to emerge.

THE PRIVATE SECTOR

One way to measure the growth of high technology company development in the Austin/San Antonio Corridor is to track employment and high technology incorporations over time. Figure 10–3 shows the incorporation of high technology companies in Austin from 1945 to 1985. Figure 10–4 shows the incorporation of high technology companies in San Antonio from 1945 to 1985. It is interesting to note that in 1983 and 1984, respectively, growth of these firms leveled off. These are manufacturing-related technology firms. They do not include service-related technology firms.

There are two other ways that we have tracked high technology company development in Austin. One is the founding or relocation of major technology-based companies. The other is an evaluation of a selected list of emerging technology-based companies. The location and homegrown development of major technology-based companies began in 1955.[2] As shown in the timetable in Figure 10–5 Austin currently has thirty-two such major firms.

Six of the companies are homegrown, and all six have had direct or indirect ties to The University of Texas at Austin. In addition, the location of the other major firms in the area was dependent on two critical elements: the presence of The University of Texas at Austin and the perception of an

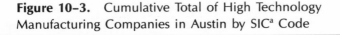

Figure 10-3. Cumulative Total of High Technology
Manufacturing Companies in Austin by SIC[a] Code

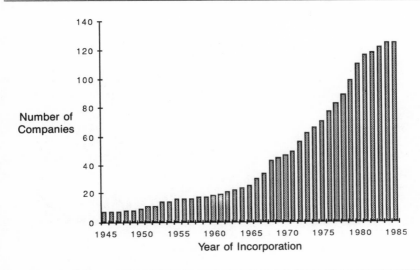

Source: 1986 Directory of Texas Manufacturers, Bureau of Business Research, Graduate
School of Business, The University of Texas at Austin.

[a]Standard industrial code.

affordable high quality of life—that is, a place with high quality-of-life factors where a company could also make a profit. Two four-year clusters are
interesting to note: 1965 through 1969 and 1980 through 1984. Major events
took place in each of these clusters. During the first, IBM located in Austin,
and during the second, MCC located in Austin.

In addition to these major firms, a second tier of small and emerging companies has been steadily increasing; 218 large and small high technology
related firms were in existence in Austin in 1986. Figure 10–6 shows their
establishment in five-year intervals from 1945 through 1985. Figure 10–7
shows the establishment of small and emerging technology related firms in
existence in Austin in 1985 in five-year intervals from 1945 through 1985.
(These charts are noncumulative; that is, they show the number of new firms
established during each five-year period.)

Of 103 small and medium-size technology-based companies in existence
in 1986, 53 or 52 percent indicated a direct or indirect tie to The University
of Texas at Austin (see Figure 10–8). These companies' founders were UT

Figure 10–4.   Cumulative Total of High Technology Manufacturing Companies in San Antonio by SIC[a] Code.

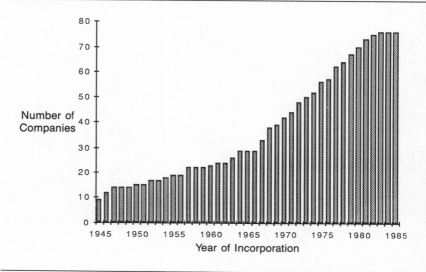

Source: 1986 Directory of Texas Manufacturers, Bureau of Business Research, Graduate School of Business, The University of Texas at Austin.

[a]Standard industrial code.

students, graduates, faculty members, or employees. These firms demonstrate an important requirement for a technopolis—the ability to generate home-grown, technology-based companies. These companies in turn have had a direct impact on job creation and economic diversification. Their tie to the university also enabled many of the companies to start their businesses with a contract that originated while they were involved in university research ac-tivities. In addition, the ability to continue their relationship in some capac-ity with the university was an influential factor in their staying in the area, along with their perception of an affordable high quality of life.

Another way to look at the tie to The University of Texas at Austin is to consider spin-out companies from selected departments and centers in the various colleges. Table 10–3 shows the type of diversity of new company development from research activities. Companies have spun out of computer sciences, physics, applied research, engineering, structural mechanics, and business. These factors can be effectively demonstrated through a case study of TRACOR, Inc., a homegrown company that is also the only *Fortune 500* company headquartered in Austin.

Figure 10–5. Major Company Relocation or Founding in Austin, 1955–86.

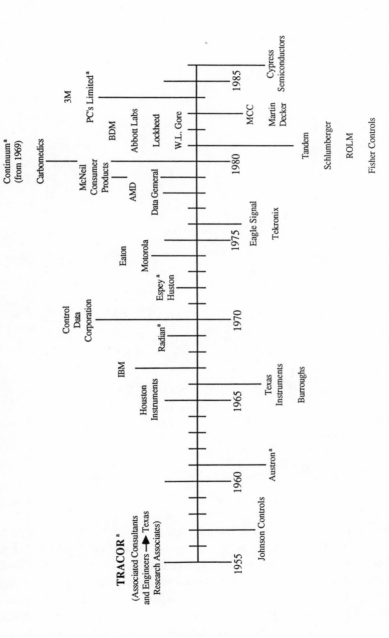

[a]Homegrown.

Figure 10-6.   Establishment of High Technology Related Firms or Branches, 1945–85.

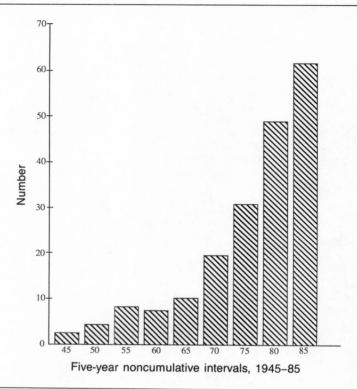

Five-year noncumulative intervals, 1945–85

Source: Directory of Austin Area High Technology Firms, 1986–87 Austin Chamber of Commerce, 1986.

TRACOR CASE

Frank McBee, the founder of Tracor, earned both bachelor's (1947) and master's (1950) degrees in mechanical engineering at The University of Texas at Austin (UT) after serving as an army air corps engineer from 1943 through 1946. After his travels in the army, McBee felt that Austin had the affordable quality of life that he wanted for himself and his family. He first worked as an instructor and then as an assistant professor in the UT Department of Mechanical Engineering. In 1950 he became the supervisor of the Mechanical Department of UT's Defense Research Laboratory (now called the Applied Research Laboratory).

Figure 10–7.   Foundings of Small and Medium-Size
Technology-Related Firms, 1945–85.

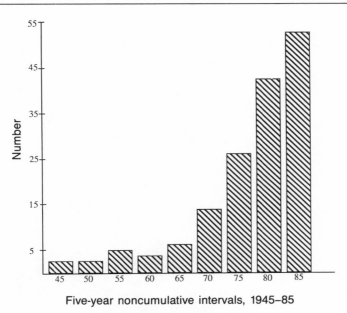

Five-year noncumulative intervals, 1945–85

Source: Directory of Austin Area High Technology Firms, 1986–87 Austin Chamber of Commerce, 1986.

In 1955, with funding of $10,000, McBee joined forces with three UT physicists and a UT-trained lawyer to form Associated Consultants and Engineers, Inc., an engineering and consulting firm. Drawing on their UT training and work experience, the four scientists focused their efforts on acoustics research. They were awarded a $5,000 contract for an industrial noise reduction project. The company's name was changed to Texas Research Associates (TRA) in 1957. During the late 1950s the four scientists taught and did research at UT while working on developing TRA. In 1962 the firm merged with a company called Textran and adopted its present name of Tracor, Inc. By this time McBee had left The University of Texas to devote himself full time to building the company.

Figure 10–9 shows that from the College of Engineering and the Defense Research Lab at The University of Texas at Austin came the educated talent to form the entrepreneurial venture of Associated Consultants and Engineers in 1955, which led to the establishment of Tracor in 1962. However, even

Figure 10–8.   Small High Tech Firms Founded with UT Connections.

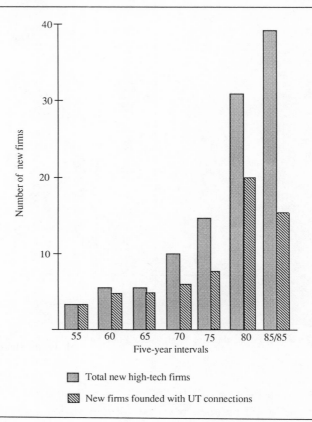

more impressive is the constant stream of entrepreneurial talent that came from Tracor itself. At least sixteen companies have spun-out of Tracor since 1962 and have located in Austin.

Figure 10–10 dramatically shows the job creation impact of Tracor and its spin-outs on the Austin area. A total of 5,467 employees were employed in these companies as of 1985. Perhaps most impressive is that some of these spin-outs have the potential of becoming *Fortune* 500 companies as their parent Tracor did. All are also capable of creating spin-outs of their own. Radian Corporation, as one example, has spun-out four companies. Most important, neither Tracor, its spin-outs, nor the jobs they created would exist without The University of Texas at Austin.

Table 10-3.   Selected UT Spin-Outs by Department.

Computer Sciences Department	*Applied Research Laboratory*
Information Research Associates	Modular Power Systems
MRI (since became a division of INTEL)	Electro-Mechanics
Statcom	National Instruments
Knowledge Engineering	TRACOR
Cole & Vansickle	*Engineering Department*
Computation Center	Mesa Instruments
Balcones Computing Co.	Geotronics Corp.
Physics Department	White Instruments
Lacoste & Romberg	Wight Engineering
Astro Mechanics	Execucom
Texas Nuclear	
Columbia Scientific, Ltd.	*Structural Mechanics*
Scientific Measurement Systems	Tekcon
Eaton Corp.	*College of Business*
Texion	ARC

Source: Ladendorf, 1982).

The private-sector association with and effect on the technopolis can be summarized as follows:

1. Companies have spun-out of The University of Texas system.
2. Major firms have been attracted and chosen to locate here for two primary reasons: access to university resources, particularly the talent pool and desire to participate in an affordable quality of life environment.
3. Employment has grown around technologically based companies.

GOVERNMENT SEGMENTS

Federal, state, and local government have also played a vital role in the development of the Austin/San Antonio Corridor as a technopolis. However, each level of government has impacted on the respective areas' economic development in different ways.

Federal government has affected the region in two key ways: through the development and operation of U.S. military bases and through federal funding for research and development activities on site and at major universities in both cities. Table 10–4 shows the impact of military bases in the Corridor. All the bases provide a general economic stimulation to the region through their employment of civilian and military personnel. For example, a San Antonio

Figure 10-9. Development of TRACOR and Its Spin-Outs, 1947–84.

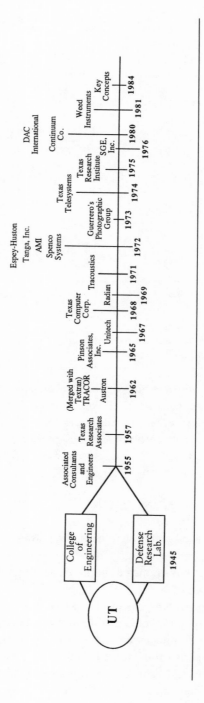

Figure 10–10. Job Creation Impact of TRACOR and Its Spin-Outs.

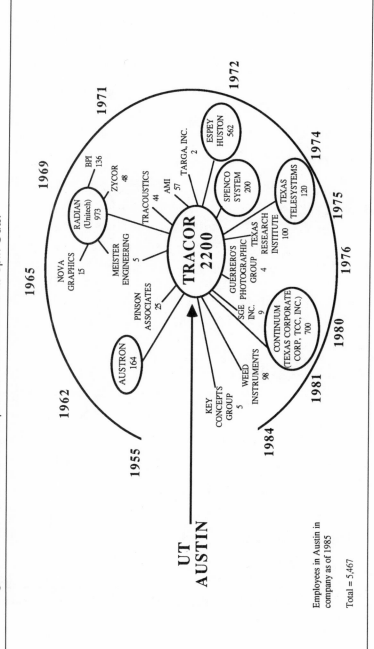

Employees in Austin in
company as of 1985

Total = 5,467

Table 10–4. Data on Military Bases in Austin/San
Antonio Corridor.

	Founded	Personnel	Annual Payroll
Austin			
Bergstrom	1942	1,000 C[a] 6,000 M[b]	$167 million
San Antonio			
Fort Sam Houston	1876	6,000 C 12,000 M	42 million
Kelly AFB	1917	15,000 C 6,000 M	
Brooks AFB	1918	600 C 2,000 M	72 million
Lackland AFB	1941	8,000 C 11,000 M	
Randolph AFB	1930	2,500 C 5,500 M	

Medical Centers, Medical Training & Research Programs

Brooke Army Medical Center
    692 beds
    200 on Gary Research Partners

Institute of Surgical Research
    $1.3 million research budget

Academy of Health Sciences
    32,000 resident students
    42,000 correspondence course students

Wilford Hall USAF Medical Center
    1,000 beds
    300 active clinical investigators

Aerospace Medical Division
    $120 million research budget

Source: San Antonio Chamber of Commerce and military bases publications.
[a]C = Civilian.
[b]M = Military.

chamber of commerce study determined that the military bases add $2.6 billion annually to the city's economy.

As noted, state government in Texas is the primary source of support for public universities, including The University of Texas at Austin, The University of Texas at San Antonio, and The University of Texas Health Science Center in San Antonio. State funding for higher education had been increasing until 1983 when Texas cut back on appropriations for higher education, just when every other state was increasing funding for education. The result was that in 1984 and 1985 the image of Texas being committed to achieving excellence in education was questioned.

Spurred by general economic slowdown and a desire to promote economic development, the 1987 state legislative session took a more proactive role. A series of new legislative proposals was presented to spur economic development and technological diversification. These include bills for business incubator support, state venture capital funding, a growth fund to spur product development, and other programs designed to assist new company development.

Although state government's primary role has been in relation to education, local government's primary role in Austin and San Antonio has focused on quality of life, competitive rate structures for such items as utilities, and infrastructure requirements. *Quality of life* carries different meanings given the subjective attributes of the issues involved. In Austin, a high quality of life has remained affordable in comparison to other technology centers. Perhaps the most dramatic statement in support of this view is the fact that MCC, which listed an affordable quality of life as one of its four main site selection criteria, decided to locate in Austin. A quality-of-life survey done at the time rated Austin as exceptional (when compared to San Diego, California; Atlanta, Georgia; and Raleigh-Durham, North Carolina) in terms of quality of primary and secondary schools, quality of parks and playgrounds, outdoor recreational opportunities, community cleanliness, and as a place to live.

Perceptions vary within any region undergoing rapid economic growth associated with a developing technopolis, and there is always the possibility that such growth will diminish the very qualities that caused the area to be attractive to high technology companies in the first place. This fine balance between a sustained quality of life and sustained economic development has been evident throughout the development of the Austin/San Antonio Corridor. One of the main reasons that Tracor located and grew in Austin and one of the main reasons that Tracor spin-outs were able to and wanted to locate in Austin was the *affordable* quality of life.

On the other hand, with each new economic development activity there was likely to be some community group that felt the loss of some, from their view central, aspect of Austin that made the city unique, desirable, and affordable. Such a list of "losses" might include more days when Barton Springs Pool, the city's best swimming location, is closed because the spring-fed pool is too full of silt from run-off at construction sites; the loss of landmarks, such as the Armadillo World Headquarters, where music greats and yet-to-be greats performed in a casual, intimate setting; and the loss of affordable land and housing. The list could go on and would vary in intensity depending on one's point of view.

Over the history of the economic development of the Austin area local government has tended to favor either the developers or the environmentalists. The issue becomes more complex because many developers are Austinites who also want to preserve what they view as Austin's quality of life. On the other hand, many environmentalists also favor some economic development. Indeed, quality of life and economic development are two sides of the same coin: Each has a vital impact on the other.

When local government supports economic development, then the development of the technopolis is more likely to increase—that is, company relocation seems to be facilitated and obstacles to development seem to diminish. On the other hand, when local government believes that quality of life is diminishing, then the development of the technopolis is inhibited—that is, company locations go to outlying cities, obstacles for development increase (such as high utility rates or slower permitting procedures), and the ability to work with diverse segments of the community declines.

One final point needs to be made concerning the quality of life issue. Although environmentalists and developers may disagree on what constitutes sensible environmental/development policy, many would agree that overall quality of life suffers most when the people who inhabit the Corridor are out of work and cannot afford to pay the costs associated with infrastructure development, housing, or such things as expanded park land or recreational opportunities.

SUPPORT GROUPS SEGMENT

Support groups have provided an important networking mechanism for the development of the technopolis. These groups take a variety of forms. Business-based groups in the Corridor relate to the emergence of specific components for high technology support in the practices of big-eight accounting firms, key law firms, major banks, and other companies. These components provide a source of expertise, even when embryonic, and a reference source for those founding and/or running technology-based enterprises.

The growth of venture capital in the Corridor provides a good example. Venture capital sources in the Corridor increased significantly in the 1980s. The growth was due primarily to two factors, one external and the other internal. Externally, changes in federal tax laws in 1979 and 1986 pertaining to capital gains encouraged investments in venture capital pools. Internally, the perception of the Corridor as an emerging technology center encouraged the development of homegrown pools. The sources of the venture capital were a few individuals knowledgable about the venture capital process as well as the major commercial banks in the area. Although funds in these pools increased, most venture capital investments continued to be made *outside* the Corridor and the state of Texas. Venture capitalists in the Corridor want a local window on technology and company development but still do not see enough good deals (that is, fast-growth company potentials) in the region.

The chamber of commerce is another important support group. It can provide a focal point for information about and support of technology-based companies. The Austin chamber of commerce, for example, played a key role in attracting IBM in 1967. It has also helped to establish other efforts to further expand the high technology network in the city. Such efforts included a highly publicized major study by Stanford Research Institute in 1983 that focused on Austin's potential as an "idea" city and one with real opportunities in specific high technology industries; expanded programs to attract and retain Japanese companies; and new organizations to broaden networks among and between technology-based organizations.

In San Antonio, the chamber of commerce, in conjunction with the mayor and city council, has proven to be a catalyst for cooperative activities to expand the south end of the technopolis. They have, for example, conducted annual economic development conferences to bring together various components of the city. They were instrumental in attracting the UT Health Science Center and UT San Antonio, and they raised the necessary private funds to ensure the creation of an engineering school.

Community groups have emerged to broaden the links and facilitate the communication process among and between technology-based organizations. The most notable development in the area was the organization of the Greater Austin/San Antonio Corridor Council in 1983. The council has provided a high-level mechanism to link key individuals and organizations from both cities. It has significantly contributed to the growing perception of the area as one region with mutually beneficial opportunities and similar problems.

Other community groups have served to try to bring together sometimes diverse and even opposing viewpoints to find common ground to address problems of mutual interest. Such groups include breakfast groups, policy-oriented groups, and special high technology groups, such as the Greater Austin Technology Business Network, which was established in late 1986, and a risk capital network system that was established at UT San Antonio in early 1987.

INFLUENCERS

Although each of the institutional segments in the Technopolis Wheel are important to the development of a technopolis, the ability to link the segments is most critical. Indeed, unless the segments are linked in a synergistic way, then the development of the technopolis slows or stops. In the Austin/San Antonio Corridor, these segments have been linked by first- and second-level influencers—key individuals who make things happen and who are able to link themselves with other influencers in each of the other segments as well as within each segment.

First-level influencers have a number of criteria in common:

1. They provide leadership in their specific segment because of their recognized success in that segment.
2. They maintain extensive personal and professional links to all or almost all the other segments.
3. They are highly educated.
4. They move in and out of the other segments with ease—that is, they are accepted and consequently help in establishing requirements for success.
5. They are perceived to have credibility by others in the other segments.

The second linkage is by second-level influencers within each segment. The second-level influencer interacts and generally has the confidence of the first-level influencer. The role and scope of the second-level influencers is to act as gatekeepers in terms of their abilities to increase or decrease flows of information to first-level influencers. They also have their own linkages to other second-level influencers in the other institutional segments. In many cases, the first-level and second-level influencers initiate new organizational arrangements to institutionalize the linkage between business, government, and academia.

Influencers seem to coalesce around key events or activities. They then play a crucial role in conception, initiation, implementation, and coordination of these events or activities. Interestingly, once an event or action is successfully managed or achieved, they help institutionalize the process so that it can function effectively without them. Consequently, an important characteristic of a technopolis is to be able to develop first-level influencers and nurture second-level influencers in all segments of the Technopolis Wheel.

Both first- and second-level influencers build extensive networks. The larger the number of influencers, the more extensive their networks, and the more they are able to interact effectively (that is, be persuasive) with all the other segments, the more rapidly the technopolis develops. Influencers play a particularly important networking role through the support groups because these groups can provide convenient opportunities to interact across all segments of the Wheel.

San Antonio can be used as an example of the role of influencers. In 1947 the San Antonio Medical Foundation was chartered by a few prominent physicians. They realized that a large amount of land was critical to the long-term success of a medical center. Consequently, they acquired over 620 acres of land in the then uninhabited western area of San Antonio. Later the acreage was increased to almost 1,000 acres. Today San Antonio has the largest medical center in the United States in terms of acreage.

Situated on this acreage as of 1987 are eight major hospitals totaling 2,893 beds, over six allied research services, and several specialized rehabilitation centers. The hub of this complex is The University of Texas Health Science Center. With more than 15,300 employees in the medical center, it is one of San Antonio's largest employers. The combined annual budget of the center's facilities is over $500 million.

Another example is the Southwest Research Institute in San Antonio. It was founded in 1947 as a not-for-profit research and development organization. The institute works at any given time on more than 1,000 engineering and physical science projects including biotechnology programs. It has currently more than 2,000 employees. Its gross revenues are just under $149 million. A Southwest Research Consortium has been formed by the Southwest Research Institute, The University of Texas Health Science Center at San Antonio, The University of Texas at San Antonio, and the Southwest Foundation for Biomedical Research. These institutions participate in cooperative projects.

SUMMARY

Although the current general economic situation in Texas has affected all the segments that make up the Austin/San Antonio Technopolis Wheel, it is possible to examine the reality of this developing technopolis based on the data and analysis in this study. The following series of tables compare Austin and San Antonio along the three dimensions for measuring the dynamics of a modern technopolis—namely, the achievement of scientific preeminence; the development and maintenance of new technologies for emerging industries; and the attraction of major technology companies and the creation of home-grown technology companies.

These tables confirm that the two ends of the Corridor complement and extend the resources of the other city on many dimensions. Table 10–5 shows the achievement of scientific preeminence in Austin and San Antonio. The segments that stand out in San Antonio are the federal government and private sectors in terms of R&D contracts and grants. In Austin the university segment is predominant. Both cities have major higher education student populations.

Table 10–5. Achievement of Scientific Preeminence.

Criteria	Austin	San Antonio
1. R&D contracts and grants		
University segment	$120 million	$ 40 million
Nonprofit research institute		150 million
Federal agency research		124 million
2. Chairs, professorships, and fellowships	470	
3. Membership in national organizations		
Nobel prize holders	2	
Membership in national academies of science and engineering	38	
4. Number of students in higher education (Public and private universities and colleges)	87,000	52,100
5. Merit Scholars	916	
6. Newer institutional relationships		
Industrial R&D consortia	MCC	
Academic/business collaboration	$25 million	
Research and engineering centers of excellence		7
7. University research and engineering centers	50	

Over half the student population in San Antonio is involved with technical training and education. In addition, San Antonio is providing unique curriculums in secondary education in health and high technology.

Table 10–6 confirms that the Corridor has been developing and maintaining new technology for emerging industries. Both cities have provided

Table 10–6. Development and Maintenance of New Technologies for Emerging Industries.

Criteria	Austin	San Antonio
1. Advanced cutting edge technologies—universities, federal agencies and research institutes	Biotechnology (plant, genetics, molecular) Semiconductor Fusion Electrochemistry New materials Theoretical physics Theoretical chemistry Aeronautical Earth sciences Computer science Artificial intelligence Computerized integrated manufacturing Vision research	Medical Instrumentation New technology products Products for use in rehabilitation and home health Biomedicine Cancer Infectious and virus-related disease Aging Burns Premature births Biotechnology (genetics, molecular) Human-centered research applications to weapons systems design and operations
Industrial R&D	AI–Expert systems Electronics Semiconductors Advanced computer research Software Biotechnology	Biotechnology Computer research Semiconductor
2. New Institutional relations Industrial R&D consortia Academic/government/ business university/ industry/university co-operative research center	MCC	Institute for Biotechnology Center for Bio Science and Technology

environments and acquired a variety of resources for advanced cutting-edge technologies. These technological areas have been developed and improved through university, federal government, and other research institutions. In addition, the private sector in both cities has been conducting industrial R&D activities in leading technological areas. New institutional relationships are emerging but need to be aggressively pursued to maintain momentum in linking new technology developments to become a mature technopolis.

Table 10–7 confirms that both cities have succeeded in attracting major technology companies and creating homegrown technology manufacturing companies. Both cities have achieved significant percentages—that is, over 5 percent, of the state's employment in high technology manufacturing, according to SIC code data. These numbers do not include additional company creation and employment in high technology services, software, and other areas. San Antonio, for example, has significant service employment in the biotechnology and health-related areas.

CASE STUDIES

The Austin/San Antonio technopolis is not an overnight success story. It has been nearly thirty years in the making. Two key events in 1983 were pivotal for the development of the technopolis. One was the decision of locating MCC

Table 10–7. Attraction of Major Technology Companies and Creation of Homegrown Technology Companies Based on SIC Codes.

	Austin	San Antonio
Attraction of major technology companies	32	13
Homegrown technology manufacturing companies, 1985		
1. By SIC codes	93	62
2. Services, software, others	NA	NA
Job creation, percent of state of Texas high technology manufacturing employment, 1985	8.1%	5.5%
Service industry employment in Medical Center, Southwest Research Institute, and Southwest Biomedical Research Institute		17,800

in Austin, Texas, and the other was the establishment of the Texas Research and Technology Foundation in San Antonio to support research and technological innovation, especially in biotechnology.

THE MCC STORY

Four states were in the final competition for the Microelectronics and Computer Technology Corporation, MCC. These competing sites represented a mature technopolis: Raleigh-Durham, North Carolina; two developing technopoleis: Austin, Texas, and Atlanta, Georgia; and an emerging technopolis: San Diego, California. These four contenders were selected from a nationwide preliminary competition that included fifty-seven major competitors from the East Coast, the West Coast, the Midwest, and the South. The four finalists were chosen for a broad range of reasons, but the primary selection criteria concerned the following:

1. Ready access to universities that are leaders in graduate-level teaching and research in microelectronics and computer sciences;
2. Good quality of life to facilitate recruitment of technical personnel (such as primary and secondary education facilities, affordable housing);
3. Easy access by air from major metropolitan areas;
4. State and local government that provides a favorable business climate; and
5. The overall cost of operating MCC.

Each of the four finalist sites met certain criteria on this list more completely than other criteria, but each of the areas had at least one major research university and each of the areas had a perceived, affordable high quality of life. Many potential candidate areas that were considered outstanding missed making the final competition because they lacked one fundamentally important criteria: a major research university. This was one item that could not be argued, as more subjective issues such as quality of life were.

Many reasons have been offered for MCC's decision to locate in Austin. Two reasons offered by observers outside of Texas that we consider myths are (1) that Texas bought MCC and (2) that Admiral Bobby Ray Inman (the founding president of MCC) had strong ties to Austin—he is a University of Texas alumni and was born in Texas—and consequently influenced the vote in favor of the area where he wanted to live, which was Austin. Although Texas did offer substantial economic incentives, this effort reflected a Texaswide fundraising program in the private sector that was viewed as an investment in the future, not an expenditure of funds that would not be recouped. In terms of the second myth (concerning Admiral Inman's ties to Austin), it could be argued that the admiral might favor San Diego more than Austin because of his family and navy ties to Southern California. In fact, the vote on the part of the MCC selection team was unanimous for locating MCC in Austin, Texas, *before* Inman cast his vote.

Based on interviews with key participants on the MCC site selection team, and in keeping with the theme of this chapter, one central issue stands above all others as the reason that MCC decided to locate in Austin: the segments of the Technopolis Wheel, especially statewide, were balanced and working. First- and/or second-level influencers in academic, business, and government organizations pulled together to propose a Texas incentive that set it apart from the other three areas. The governor of Texas and his high-level representatives coordinated, organized, and worked with the regents and high-level administrators of The University of Texas at Austin and Texas A&M University to find ways to demonstrate strong statewide support (that is, in terms of endowed chairs, professorships, student support) for the Departments of Electrical Engineering and Computer Science. A statewide funding effort with a goal of $23 million was initiated and carried out by business leaders from Dallas to Houston to Amarillo to fund these academic incentives as well as other MCC inducements. Local banks and the Austin business community put a package together to subsidize mortgage loans for potential MCC employees to the amount of $20 million. Dedicated attention was given to other incentives such as offering staff and resources to help the spouses of MCC employees find suitable employment in the Austin area.

A pointed example of the need for a balanced and working Technopolis Wheel is provided by the volunteer team of academic/business/community leaders who met daily in a "war room" and worked for two intense weeks to craft and finalize the Texas incentive for MCC. These prominent leaders and their support staff came from the following professions: state government, law, public relations, developers, industry, consulting firms, and the University of Texas and Texas A&M. Individually these team members represented a range of talents and professional skills. Together they had strong ties with and a working knowledge of all segments within the Technopolis Wheel: The University of Texas and Texas A&M, state and local government, and the state and local business community. The mayor and staff of San Antonio helped initiate the spirit of cooperation by joining the Austin effort when it became clear that there would be only one Texas finalist, Austin.

For the two weeks between the preliminary and final selection decision, there was a remarkable spirit of cooperation where Texans from each of the segments of the Wheel gave freely of their time and talents to win MCC for Texas and Austin. They were driven by the spirit of competition with the other three finalist sites and the vision of what MCC meant for the Austin/San Antonio Corridor, the state of Texas, and for the competitive advantage of U.S. high technology industry. The spirit of a team of prominent individuals working together for the common good was so strong that arguments over parochial issues were put aside.

Although the other sites also made impressive offers to MCC in many different ways, none of them came close to displaying Texas's cooperative spirit in the MCC bid. It is interesting to note that while for Texas the synergy was there in an intense way in 1983, by 1987 it was less apparent for several reasons: a general statewide economic slowdown, questions over adequate state government funding for education, and concerns about local government inhibiting high tech development and economic growth in general.

Retrospectively, in a sense, Austin the winner in the MCC competition, lost, and the cities that lost became the winners. When Austin won MCC, government, academic, and business leaders in other states with high tech fever saw or feared that Texas and its universities (principally The University of Texas at Austin and Texas A&M) had the momentum to outdistance their efforts in developing academic institutions of excellence necessary for a technopolis. They envisioned their flagship universities and states being stripped of some of their most valued resources: outstanding professors and their students and entrepreneurial business and community leaders. They envisioned other companies following the lead of Lockheed, 3M, and Motorola either relocating old divisions or moving new divisions to the developing technopolis of Austin, Texas. Their fears were unfounded. Shortly after MCC selected Austin, Texas became the only state to actually cut appropriations for higher education and for a variety of reasons the growth of the Austin/San Antonio technopolis slowed.

By 1987 in Austin there was a loss of synergy between the private sector and local government. Some would argue that the development of the area was moving too fast and that this was reflected in soaring land prices and a declining quality of life. On the other hand, others would argue that the city council had, either by action (such as increased electric rates, a web of building permits, or time and effort spent on countless meetings on minor issues) or nonaction (such as making no decisions on important projects such as building a new airport or a new convention center) played too great a role in slowing economic development.

An important effect of the state's approach to funding higher education and of changing Austin City Council policies concerning economic development was that academic talent and business leaders became hesitant to locate in a region where the rules of operation were constantly changing. Indeed, even some Austin-based technology businesses are being tempted to relocate to more accommodating, stable areas within and outside of Texas.

The MCC competition demonstrated that for the institutional segments of the Technopolis Wheel to cooperate, technopolis participants (that is, academic/business/community leaders) cannot become so consumed with the parochial interests that they lose the long-term vision. It is the vision that binds the participants to a policy that makes the activity attractive to others and that sustains long-term, steady development of the technopolis.

THE DEVELOPMENT OF BIOTECHNOLOGY
IN SAN ANTONIO

In 1983 Mayor Henry Cisneros launched "Target '90. Goals for San Antonio to Build a Greater City," a long-range, communitywide planning project for the City of San Antonio. Members of the Target '90 Commission included first- and second-level influencers from government, business, and academia. The Target '90 process helped build a remarkable consensus about the priorities for and direction of the city through 1990. The final report targeted 177 specific action initiatives to build the future of San Antonio. A key section of the report focused on developing San Antonio as a key center for biotechnology in the future.

The plan focused on expanding on extensive university, private/profit, non-profit, and military research entities already located in San Antonio, including Southwest Foundation for Biomedical Research, Southwest Research Institute (the third-largest private research institute in the nation), The University of Texas Health Science Center, the U.S. Air Force's School of Aerospace Medicine at Brooks Air Force Base, the U.S. Army's Health Service Command and Brooke Army Medical Center at Fort Sam Houston.

In March 1984 the UT Health Science Center in San Antonio with encouragement of the mayor's office succeeded in winning a $75,000 planning grant from the National Science Foundation to develop a University-Industry Cooperative Research Center for Biosciences and Technology. The UICRC grant focused on developing a partnership between the Health Science Center and industry nationwide. It brought credibility and prestige to the effort for building the Research Center. The focus from the start was to link basic research with opportunities for local economic development and jobs.

At the same time, the HSC had proposed to the community through Target '90 an Institute for Biotechnology. The institutional segments of the

Technopolis Wheel then began to focus momentum for a major research presence in biotechnology through the creation of a new institution, the Texas Research and Technology Foundation, that was formally established in 1985. This foundation is a nonprofit economic development organization to support scientific research and technological innovation. Its objective is to promote and build on San Antonio's existing and extensive technology base. The board of the foundation includes first- and second-level influencers from the city's civic, business, academic, scientific, and professional communities. The foundation's method seeks to link the university, industry, and government. In this regard, it has done the following:

1. Established the Texas Research Park with a private donation of 1,500 acres;
2. Provided fifty acres of land to the Institute for Biotechnology to be developed in collaboration with The University of Texas Board of Regents and UT Health Science Center;
3. Created an Invention and Investment Institute to facilitate a technology venture or incubator center and venture funding; and
4. Supported an Institute of Applied Sciences to be developed in collaboration with the Southwest Research Institute to transfer basic research discoveries and the know-how of local scientists to private companies and government agencies.

From 1983 to 1987 interest in and programs for biotechnology increased. Ten of thirteen public and private colleges and universities in the city now offer curriculum in the biomedical/biotechnology areas. The most notable example is The University of Texas at San Antonio, which has established a nonthesis master's degree program in biotechnology. It was the first of its kind in Texas and is one of a handful of such programs nationally. In June 1986 the growth of the classes in engineering and biotechnology warranted the authorization by the UT System Board of Regents of a $27.9 million engineering/biotechnology building at the university. (Even area high schools have developed programs in high technology and biomedical instruction.)

In April 1986 the UT System Board of Regents formally approved locating the UT Institute for Biotechnology in the Texas Research Park. By that time, Concord Oil Company had donated 1,500 acres to the Texas Research and Technology Foundation to develop the research park with the institute as its first project. The foundation's land committee, which secured the donation, was composed of top influencers in the city. In addition, the foundation planned to raise $20 million in cash for building construction, equipment, and the endowment of the Institute. In January 1987 the UT System Board of Regents approved $10 million to be matched by the private sector to build a biotechnology research building at the UT Health Science Center.

Today, San Antonio can claim a new research park, a biotechnology institute that links the university with industry, and a growing critical mass in the biotechnology industry. One important indication of that mass is the eighteen biotechnology companies that are now operating in San Antonio.

CONCLUSION

A number of key points emerged from the Austin/San Antonio Corridor study:

1. The research university has played a pivotal role in the development of the technopolis by

   Achieving scientific preeminence,
   Creating, developing, and maintaining the new technologies for emerging industries,
   Educating and training the required work force and professions for economic development through technology,
   Attracting large technology companies,
   Promoting the development of home-grown technologies, and
   Contributing to improved quality of life and culture.

2. Local government has had a significant impact, both positively and negatively, on company formation and relocation, largely from what it has chosen to do or not to do in terms of quality of life, competitive rate structures, and infrastructure.

3. State government has had a significant impact, both positively and negatively, on the development of the technopolis through what it has chosen to do or not to do for education, especially in the areas of making and keeping long-term commitments to fund R&D, faculty salaries, student support, and related educational development activities.

4. The federal government has played an indirect but supportive role largely through its allocation of research and development monies to universities, on-site R&D programs, and defense-related activities.

5. Continuity in local, state, and federal government policies has an important effect on maintaining the momentum in the growth of a technopolis.

6. Large technology companies have played a catalytic role in the expansion of the technopolis by

    Maintaining relationships with major research universities,
    Becoming a source of talent for the development of new companies,
    Contributing to job creation and an economic base that can support an affordable quality of life.

7. Small technology companies have been increasing steadily in number and size in the area. They have helped in

    Commercializing technologies,
    Diversifying and broadening the economic base of the area,
    Contributing to job creation,
    Spinning companies out of the university and other research institutes, and
    Providing opportunities for venture capital investment.

8. Influencers have provided vision, communication, and trust for developing consensus for economic development and technology diversification, especially through their ability to network with other individuals and institutions.

9. Consensus among and between segments is essential for the growth and expansion of the technopolis.

10. Affordable quality of life, while subjective and hard to measure, is important in the development of a technopolis. It can be a major source of friction between advocates and adversaries of the growth of the technopolis.

11. The very success of the development of a technopolis can lead to greed and many dissatisfactions. The result can be a shattering of the consensus that originally made the technopolis possible.

The Austin/San Antonio Corridor is a developing technopolis with promise. The area has been achieving scientific preeminence, developing and maintaining new technologies for emerging industries, and attracting large technology firms while creating homegrown technology companies. It still has a way to go before reaching maturity.

In conclusion, there needs to be a broader vision of the future. A future that is not lost in local and state economic setbacks or interminable resolutions about affordable quality of life. A future that comes from utilizing the

Corridor's most important resource—its intellectual resources found within its institutional segments. A future that provides a vision for effectively linking government, business, and academia.

NOTES

1. An important point needs to be made about higher education in general in the Austin/San Antonio Corridor. Other universities have also provided research, teaching, and training that contributed to the development of this technopolis. Some of these include Southwestern University in Georgetown (twenty miles north of Austin); in Austin, St. Edward's University, Austin Community College, and Concordia College; Southwest Texas State University in San Marcos (twenty miles south of Austin); in San Antonio, Trinity University, St. Mary's University, and Our Lady of the Lake College. These universities have over 139,000 students. The other flagship university in the state, Texas A&M University in College Station, 100 miles northeast of Austin, also affected the Corridor through its major research activities. There is some speculation that the technopolis of the Corridor may eventually form a crescent by looping from Austin to College Station.

2. By *major technology-based companies,* we mean headquarters and branches of *Fortune* 500 companies and/or those companies with annual revenues or annual R&D budgets of over $50 million and/or those companies with over 450 employees in Austin.

SELECTED BIBLIOGRAPHY ON THE AUSTIN/ SAN ANTONIO CORRIDOR

Academe-Industry Linkage Forum Proceedings. 1984. Austin, Tex.: United San Antonio and The San Antonio Medical Foundation.

"Austin Chamber of Commerce 1983 Annual Report." 1984. *Austin Magazine* (January).

"Austin Chamber of Commerce 1984 Annual Report." 1985. *Austin Magazine* (January).

"Austin Chamber of Commerce 1985 Annual Report." 1986. *Austin Magazine* (January).

"Austin Community Profile 1986–1987." 1986. Austin, Tex.: Austin Chamber of Commerce.

"Austin Economic Review—Living Up to Expectations." 1981. *Austin Magazine* September, pp. 5a–16a.

"Austinite of the Year: Frank McBee." 1987. *Austin Magazine* (January): 30–38.

Brooke, Ann. 1980. "Plugging into Austin's Booming Electronics Industry." *Austin Magazine* (October): 39–47.

"Creating an Opportunity Economy." 1985. Final report of SRI Project 7799 prepared by SRI International for the Austin Chamber of Commerce. April.

Defense Contract Administration Services Region News Release on Defense Contracting. 1986. January 15.

"Directory of Austin Area High Technology Firms 1986–1987." 1986. (Austin, Tex.: Austin Chamber of Commerce).

Downing, Diane E. 1983. "Thinking of the Future: The Promise of MCC." *Austin Magazine* (August): 105–10.

———. 1984. "Casting a High Tech Spotlight." *Austin Magazine* (January): 32–35.

"Economic Growth and Investment in Higher Education: Summary and Selected Bibliography." 1987. Bureau of Business Research, The University of Texas at Austin.

"The Economic Review and Forecast." 1985. Supplement to *Austin Magazine* (October).

"The Economic Review and Forecast." 1986. Supplement to *Austin Magazine* (October).

"The Emerging Economic Base and Local Development Policy Issues in the Austin-San Antonio Corridor." 1985. Lyndon B. Johnson School of Public Affairs, The University of Texas at Austin. Policy Research Project Report Number 71.

English, Sarah Jane. 1983. "A Vintage Friendship." *Austin Magazine* (September): 46–52.

Farley, Josh, and Norman J. Glickman. 1985. "R&D as an Economic Development Strategy: The Microelectronic and Computer Technology Corporation Comes to Austin, Texas." Austin, Tex.: Department of Planning and Growth Management, City of Austin.

"The Future of the Austin-San Antonio Corridor." 1985. Proceedings of *The Future of the Austin-San Antonio Corridor*. Austin, Tex.: The Greater Austin-San Antonio Corridor Council, Inc., San Marcos, Tex. and The Population Resource Center, New York, New York.

"Growing Research Activity Catches On Here." 1961. *Austin in Action* (October): 8–19.

"High-Tech Companies Team up in the R&D Race." 1983. *Businessweek* (August 15): 94–95.

Ladendorf, Kirk. 1983. "The Anticipation Ends: MCC Moves In, but Is Still Feeling Some Growing Pains." *The Austin American-Statesman* (September 12).

Lineback, J. Robert. "Texas Cash Fuels Electronics Boom." 1983. *Electronics* (June 16): 95–96.

Long, Mary Kathryn. 1982. "Austin, Texas: An Analysis of Business Growth and Development." MBA Professional Report, University of Texas at Austin.

McBee, Frank. W., Jr. 1986. "What's Best for Texas and Austin." Speech given at Annual Meeting of the Austin Chamber of Commerce, Austin, Tex., January 21.

———. 1987. "Education & the Business Community." Special Briefing to Friends of Higher Education Conference, Austin, Tex., February 11.

"MCC Impact Assessment: Economic, Demographic, and Land Use Impacts of the Microelectrics and Computer Technology Corporation." 1984. Joint Research Project of the Graduate Program in Community and Regional Planning, The University of Texas at Austin and the Department of Planning and Growth Management, City of Austin.

"1982 Economic Review." 1982. *Austin Magazine* (October).

"1984: The Year in Review, The Year Ahead." 1984. *Austin Magazine* (October).

"1987 Target Market Program Economic Development." 1986. Austin, Tex.: Austin Chamber of Commerce.

Provine, John. 1984. "The Makers of High-Tech." *Austin Magazine* (November): 8a–28a.

"Quality of Life: Austin Trends 1970–1990." 1984. Research report prepared by students as part of classwork under Dr. Dowell Myers, Community and Regional Planning Program, School of Architecture, University of Texas at Austin. June.

"Quality of Life Indicators for Austin." 1986. Report by the Quality of Life Advisory Committee to the Quality of Life Division, Austin Chamber of Commerce. February.

"Research Venture Gets U.S. Approval on Computer Work." 1985. *Wall Street Journal,* March 5.

"San Antonio: On the Fast Track for Biomedical/Biotech Growth." 1986. Department of Economic and Employment Development, The City of San Antonio.

"San Antonio's Electronic Industry." 1987. Department of Economic & Employment Development, The City of San Antonio. February.

Segal Quince Wicksteed, Economic and Management Consultants. 1987. "University-Industry Research Links and Local Economic Development: Food for Thought from Cambridge and Elsewhere. Unpublished Study. February.

Shahin, Jim. "MCC: Keeping the Promise? Texas Pins Its Hopes on a New Deal." 1985. *Texas Technologies* (November): 9–17.

————. 1985. "Courting MCC." *Texas Technologies* (November): 11.

"Statistical Handbook 1986–1987." 1986. The University of Texas at Austin, Office of Institutional Studies.

Susskind, Hal. 1980. "It Has a Lot of Plus Factors, But How Recession-Proof Is Austin's Economy?" *Austin Magazine* (June): 3a–16a.

Szilagyi, Pete. 1980. "The Miracle Chips." *Austin Magazine* (March): 27–30.

"Technology Resource Inventory Volume One: Biosciences in San Antonio." 1986. Texas Research and Technology Foundation. June.

"Telecity. San Antonio: The Telecommunications Advantage." 1986. San Antonio Economic Development Foundation.

Tempest, Rone. 1984. "Texas Pays High Price for Luring High-Tech Venture." *Los Angeles Times,* February 21.

"United We Stand. . . ." 1975. *High-Tech Marketing* (April): 70–71.

GENERAL SELECTED BIBLIOGRAPHY

Allen, David N., and Victor Levine. 1986. *Nurturing Advanced Technology Enterprises: Emerging Issues in State and Local Economic Development Policy.* New York: Praeger Publishers.

Bates, James Vance. 1983. "Decisional Factors in Corporate Site Location." MBA Professional Report, University of Texas at Austin.

de Gennaro, Nat. "Economic Impacts of the University of Arizona." 1987. Division of Economic and Business Research, College of Business and Public Administration, The University of Arizona.

Harrigan, Kathryn Rudie. 1985. *Strategies for Joint Ventures.* Lexington, Mass.: Lexington Books.

Hisrich, Robert D., ed. 1986. *Entrepreneurship, Intrapreneurship, and Venture Capital.* Lexington, Mass.: Lexington Books.

Konecci, Eugene B., and Robert Lawrence Kuhn, eds. 1985. *Technology Venturing.* New York: Praeger Publishers.

Konecci, Eugene B. et al. 1986. "Commercializing Technology Resources for Competitive Advantage." IC^2 Institute, The University of Texas at Austin.

Krzyzowski, M. et al. 1982. "Developing High Technology Industry." In *Michigan's Fiscal and Economic Structure,* edited by H. Brazer and D. Laren. Ann Arbor: University of Michigan Press.

Larsen, Judith K. et al. 1987. "Industry-University Technology Transfer in Microelectronics." Cognos Associates and Program in Information Management, School of Public Affairs, Arizona State University.

Ouchi, William. 1984. *The M-Form Society: How American Teamwork Can Recapture the Competitive Edge.* Menlo Park, Calif.: Addison-Wesley.

"The Quality of Life in Four American Cities." 1983. Special market research study for the Governor of Texas' MCC Task Force. Shipley & Associates, Inc. under direction of Neal Spelce Communications.

Rogers, Everett M., and Judith K. Larsen. 1984. *Silicon Fever: Growth of High-Technology Culture.* New York: Basic Books.

Ryans, John K., and William L. Shanklin. 1986. *Guide to Marketing for Economic Development.* Columbus, Ohio: Publishing Horizons.

Segal Quince Wicksteed, Economic and Management Consultants. 1987. "University-Industry Research Links and Local Economic Development: Food for Thought from Cambridge and Elsewhere." Unpublished study. February.

Simison, Robert L. "Quiet War Rages on Technology's Front Line." 1985. *Wall Street Journal,* December 24, p. 4.

Watkins, Charles B. 1985. "Technology Transfer from University Research in Regional Development Strategies." *Journal of Technology Transfer,* pp. 10–11.

Chapter 11

HIGH TECHNOLOGY DEVELOPMENT IN THE PHOENIX AREA
Taming the Desert
Rolf T. Wigand

Since the mid-1950s when Motorola established its microelectronics facilities in Phoenix, Arizona has benefited from the growth of high technology industry. High technology manufacturing (computers, electronic components, aerospace, communications, and scientific instrumentation) accounts for almost 50 percent of all manufacturing jobs in Arizona, comparing to a 15 percent national average. The state, in 1987, has 313 high technology firms, up 33 percent from 225 firms in 1975, including an electronics firm founded by Hopi Indians (Webster 1987). The value of shipments exceeds $5 billion annually (Arizona Office of Economic Planning and Development 1984).

Some 85,000 workers currently are employed in high technology companies in Arizona, a growth of more than double the 42,000 employed in 1975. The state's fourteen largest firms employ 78 percent of Arizona's high technology work force. Metropolitan Phoenix accounts for 79 percent of the state's high technology jobs, and Tucson accounts for 20 percent. Arizona has experienced a high technology employment growth rate of about 6 percent annually over the past few years, while U.S. high technology employment has grown by 2.4 percent annually. High technology employment represents 7 percent of total employment in Arizona and 3 percent nationally (Price and Hodge 1984). The largest high technology employer is Motorola with about 22,000 employees.

The continued dominance of high technology manufacturing in Arizona's industrial mix has led to a more stable state economy over the last ten years.

Between 1975 and 1979 numerous new firms have moved to Arizona, including Digital Equipment, IBM, Intel, GTE, and Gould, and many firms made major improvements and expansions within their existing Arizona facilities, including Honeywell, Garrett, and Sperry. High technology employment's share of overall manufacturing employment has resulted in fewer job losses during recessionary periods. Arizona semiconductor firms, for example, seemed to view the recent semiconductor slowdown as an inventory correction. There were relatively few layoffs, and major employers tended to use four-day workweeks, hiring freezes, overtime, and cost-cutting measures. In addition, Arizona's major employers are producers of custom, rather than commodity, chips and were consequently somewhat sheltered from the overbooking phenomenon in 1983 through 1985. Those firms with contracts from the generally more stable government electronics industry did economically better than those firms not involved in this branch of the industry.

Arizona's growth rate is consistently among the fastest-growing in the country. With nearly 3 million people, the state's 39 percent growth from 1973 to 1983 placed it fourth nationally in rate of population growth. Nationally the Phoenix metropolitan area is ranked second in population growth over the last two years. Almost 70 percent of the Arizona population is located in the Phoenix metropolitan area. Phoenix proper is the nation's ninth-largest city. Arizona ranked fifth in growth of jobs and sixth in growth of income over this period. These growth figures are even more astounding when comparing Arizona's overall employment growth with that of Western Europe. Over the last fifteen years Arizona has added more jobs to its economy than France, Germany, Great Britain, Italy, and Sweden combined. More specifically, Arizona has added more jobs to its economy than all of Western Europe during this time period. With the exception of Belgium, the Netherlands, and Portugal—countries for which data were not readily available—Western Europe actually had an overall job decline over the last fifteen years.

Net in-migration, averaging well over 40,000 persons per year, is the leading contributor to population growth, with the rate of natural increase running at about 30,000 per year (University of Arizona College of Business and Public Administration 1984). The state's labor force consists of about 1.4 million workers (Arizona Department of Economic Security 1984). Some 880,000 workers resided in Maricopa County where Phoenix is located, 257,000 lived in Pima County, and the rest were spread throughout the state (Valley National Bank 1984). In 1984 employment in Arizona grew by over 8 percent, personal income by 13 percent, and retail sales by 12 percent (Vest 1985).

Arizona was the fastest-growing state in the nation from 1975 to 1985 in terms of employment and personal income. In terms of population growth,

Arizona ranked third in the nation, behind Alaska and Nevada. This performance ranked the state first in the Four Corners Region in all three aspects of economic growth. Since the recovery began in November 1982 the state's employment has grown by almost 32 percent versus 13.2 percent nationally, again making Arizona the fastest-growing state in the United States.

A major key to Arizona's success is the employment mix. The goods-producing sector (manufacturing, mining, construction) accounts for about 23 percent of total employment. Manufacturing accounts for almost 14 percent of employment, with half of the manufacturing jobs being in high technology. This compares favorably to the national average of only 12 percent of manufacturing employment being related to high technology. Even within Arizona's high technology component, there is a broad diversity between computers, components, and aerospace. As already indicated, this diversity has allowed the state to do well even though semiconductors have been in trouble for the last two to three years. The goods producing sector in Arizona has created 77,800 jobs, a 34 percent gain, since the beginning of the November 1982 recovery (Valley National Bank 1986: 2).

Although migration to the West may be slowing, Arizona's rapid growth is expected to continue and may exceed other Sunbelt states, due principally to an influx of Midwesterners (Hall 1985). Areas of rapid growth also seem to experience higher labor force turnover rates (Rogerson and Plane 1985).

Most high technology firms in Arizona are located in the Phoenix and Tucson areas. Although some high technology companies are found in Flagstaff and a few support firms are scattered around the state, most activity is located in these two largest urban areas. There are some indicators that a high technology Phoenix-Tucson corridor may come about as several high technology firms have located their facilities in Casa Grande.

HISTORY OF ARIZONA'S ROLE
IN HIGH TECHNOLOGY

High technology industry and government work reasonably well together in Arizona. Industry is included in issue-oriented dialogues with government, and government officials seem aware of, and even sympathetic to, industry needs.

Arizona's recent development strategy, formulated in 1983 under Governor Babbitt, specifically addresses high technology industry. The report "Arizona Horizons: A Strategy for Future Economic Growth" (Arizona Office of Economic Planning and Development 1983) makes the following general recommendations:

First, this strategy must put a great deal of emphasis on the high technology future of Arizona. . . . Small business vitalization must be the second major focus . . . ensuring that the optimal economic potential of all areas of the state is recognized and supported. . . . Finally, strategies for the promotion of technological innovation, business development and balanced economic growth are incomplete if they do not directly address the critical need for education and manpower training.

The following specific recommendations are made in the report:

1. Research and development

    Increase research funding and support;
    Set up a Council on Science and Technology;
    Develop a long-term state telecommunications policy;
    Review the Arizona Board of Regents' patent policy and sponsored research policies;
    Explore formal mechanisms to promote the transfer of technology between the state universities and businesses in the state as well as between the businesses themselves.

2. Capital availability

    Improve administration of state securities law;
    Develop relationships with the venture capital community nationwide;
    Encourage formation of Small Business Investment Companies;
    Authorize state employee pension funds to invest in professionally managed venture capital partnerships or corporations.

3. Entrepreneurial and small business support

    Create a governor's task force to survey approximately fifty companies;
    Establish a network of public/private assistance teams to provide preventure development assistance to entrepreneurs and investors.

4. Education and manpower training

    Continue to establish centers of excellence in the state's universities that nurture and support the development of high technology;
    Improve what is taught and learned in Arizona school systems;
    Increase teacher salaries;
    Evaluate teacher education programs and certification standards;

Encourage community colleges to continue to take a strong leadership
role in vocational technical education;
Support competency-based education.

5. Additional general recommendations

Recognize the importance of Arizona's environment and quality of life
as a component of economic growth;
Promote cooperative development efforts among communities;
Encourage public/private partnerships.

Several studies have been conducted by the Arizona Office of Economic
Planning and Development (OEPAD)—now the Arizona Department of
Commerce—to define the support industry needed by high technology
development. These reports identify opportunities created by high technology
industries, especially for new service and support industry. This office has
established a Small Business Development Corporation to assist in funding
businesses. By encouraging companies to start-up or expand, Arizona not
only assists industry already in the state but also increases the multiplier effect. Thus, high technology industry may also effect the low tech portions
of the state's economy as new products and procedures increase industry efficiency and reduce costs.

As high technology industry develops in Arizona, state government can
play an important role in assisting the maintenance of a balanced economy.
Arizona's early development was primarily resource based, the famous four
Cs—that is, copper, cotton, citrus, and cattle—and largely controlled by forces
and institutions beyond the state's boundaries. This no longer is the case.
In recent years, Arizona has been able to monitor its economic development
and guide its growth in directions that anticipate future problems.

Arizona has a commitment to stimulating further growth in its high
technology sector. Yet this growth does not occur without costs; there are
potential negative impacts as well. A major factor is the volatility of an industry based largely on innovation. Arizona's experience with the electronics
industry shows that there are cycles associated with supply and demand and
new products that make the state's economy and work force vulnerable to
sharp fluctuations.

Public health and safety issues related to high technology industry have
been acknowledged. Environmental pollution and hazardous materials
associated with the high technology industry are being addressed. According
to state officials, Arizona has adequate environmental protection laws covering

hazardous wastes and ground and surface waters (Stanton 1984). Recent publicity suggests, however, that the funding and staffing necessary to adequately administer those laws is not present.

GROWTH OF HIGH TECHNOLOGY INDUSTRY IN ARIZONA

The source of most high technology growth in Arizona is the expansion of existing companies to new facilities or the relocation of national corporations to Arizona. Companies already in Arizona usually expand to adjacent sites, especially if their needs are met and the labor market is not saturated by them or their competitors. Although corporate expansion generally does not receive a great amount of attention, it represents an important source of growth and serves as an important indicator of the overall economic stability of an area.

Another source of high technology industrial growth is a local start-up or spin-off company, founded by individuals who previously worked for other companies in the area. The general pattern for locating spin-off companies is agglomeration, so that once a critical mass of companies in high technology industry is located in one area, new companies in that industry will concentrate in areas where the original agglomeration occurred. Such was the case with semiconductor companies in Silicon Valley.

A recent source of high technology growth with significant potential is the location of international corporations in Arizona. This is similar to new facility location for any company, but there are significant differences for international corporations. One troublesome implication is the tendency for these companies to initially develop manufacturing and assembly activity in Arizona, and later to move them from the United States to offshore locations, attracted primarily by lower labor costs elsewhere. To the extent that this activity goes to northwest Mexico, it influences Arizona and the trade ties between the two regions. There are several key factors that support high technology development in Arizona: (1) an environment favorable to business, (2) a quality work force, (3) appropriate infrastructure, and (4) a research and development environment.

ENVIRONMENT FAVORABLE TO BUSINESS

A viable business environment is essential to the attraction, expansion, and retention of high technology industry. Factors influencing the business environment include taxes and governmental regulations, the availability of appropriate financing, a positive work ethic, and a favorable relationship between business and the community. A realistic and consistent state development strategy, with an understanding of the area's current development status and plans for the future, also is important. Because a corporation operates within the area's economy, the overall prosperity of the region is an important aspect of the corporation's ability to compete in its industry.

A competitive tax climate is critical. Tax rates must be realistic to meet the state's needs and the corporation's ability to pay while remaining competitive in its industry. Income and property taxes, unemployment and industrial accident insurance rates, and the procedures of taxing authorities all must be reasonable, or companies will go out of business or be driven from the state. Most corporations are well informed about comparative tax rates. Regions that are out of line or impose additional taxes or requirements, such as unitary taxes, suffer as a result.

Reasonable, consistent regulations are important to the business climate in Arizona. Companies are reluctant to locate or to remain in areas with overly complex, burdensome regulations or permit processes. Problems with regulations add to the cost of doing business, increase the time and effort required for business operations, and add to a company's uncertainty. Examples of regulatory problems include the interpretation and application of environmental regulations, occupational safety and health standards, and municipal standards.

The Arizona legislature is presently discussing bipartisan bills that could turn almost all counties into enterprise zones to encourage industrial development. Areas designated as enterprise zones would receive several tax breaks to lure new business and industry, including a 5 percent exemption from sales tax on prime contractors, a reduced transaction privilege tax, an income tax credit of $2,500 for each additional employee added to a firm's payroll, and others.

High technology industry also has specific needs from the business community, especially the banking and finance industry. From lending policies that reflect an understanding of industry conditions and practices to venture capital availability, the financial community must be responsive if it is to capture its share of business.

Two editors of *Inc.* magazine stated that economic growth depends more on quality of life and business/government cooperation than on corporate relocations (Noyes 1987). *Inc.* ranked in its April 1987 issue 150 cities based on high number of jobs generated, business start-ups, and young companies enjoying high growth rates. Phoenix was rated as the third most entrepreneurial city in the nation—after Austin, Texas, (first) and Orlando, Florida, (second)—and Tucson, Arizona, (fifth) (Noyes 1987).

QUALITY WORK FORCE

The availability of a quality work force has become basic for the development of high technology industry. The ability to innovate and the ability to produce a product in a cost-effective manner both depend on quality workers. High technology companies locate in areas with a concentration of similar companies, primarily to take advantage of available labor force and of infrastructure services. This clustering produces an economy of scale in training and creates opportunities for support industries devoted to labor force preparation.

A community's ability to sustain a quality work force is contingent on the area's desirability as a place to live. The physical setting, natural beauty, and climate are factors, along with cultural activities, outdoor recreation opportunities, and its acceptance of people new to the area. Many high technology employees can find work anywhere, so quality-of-life issues are especially important for this industry, and many of these expectations are met in Arizona. Recent attention of studies conducted by Neal Peirce and Associates, Paul Eggers, and others have brought attention to increased dissatisfaction with various quality of life aspects in the Phoenix area. Similar concerns were also reported in a *New York Times* story (Lindsey 1987).

The competitive cost of living is also important to attracting people to Arizona. The availability and cost of housing in various price ranges is important, as are food, energy, transportation costs, and personal taxes. Phoenix led the nation in housing starts in 1984 with units available in all price ranges and types (Mountain West Research–Southwest, Inc. 1984). The 1986 median new home value was $91,900, with sales substantially increasing in the over $100,000 price range; the 1986 median price for a resale home was $79,000 (Cole 1987).

A good educational system from preschool to postgraduate programs also has been critical to attracting people to Arizona. Arizona ranks twenty-third nationally in educational appropriations with $116 per capita. The cost per pupil in the regular grades K through 12 educational program is $1,788. About 589,000 students were enrolled in 879 public and 354 private schools with 26,200 teachers. Some 82,000 students were enrolled in Arizona's three universities, with 116,500 students enrolled at 15 community colleges (Arizona Department of Economic Security 1984).

APPROPRIATE INFRASTRUCTURE

Infrastructure includes the services necessary to connect a company at a specific location to the rest of the world. Although companies do not choose a location because of infrastructure availability, the lack of needed infrastructure may prevent a company from selecting an otherwise desirable location.

High technology industry has significant infrastructure requirements. For example, certain companies need 3 million gallons of water low in total dissolved solids for daily industrial process use. Arizona is perceived by many as a desert state with limited groundwater overdrafting. Some experts claim that there is no water-related reason not to locate in Arizona (Steiner 1985). Others say that although water resources need careful management, Arizona's water resources are adequate to meet foreseeable municipal and industrial needs (McNulty and Woodard 1984).

Sewer systems face growth-related needs. After industrial use, most water returns to a community's sewer system and treatment plant. Due to cost and treatment plant capacity considerations, many of the larger new developments in the Phoenix area now are building systems that treat sewage on-site and use the effluent for landscape and golf course irrigation.

Increasingly, high technology companies require a large supply of reliable, high-quality electric power. Any interruption, no matter how brief, can cause the loss of millions of dollars worth of work in process; voltage fluctuations can result in similar costly losses. Adequate supplies of electrical energy are available in Arizona. Although distribution capabilities vary, they are generally adequate to support considerable load growth. Power costs in Arizona are in the midrange of costs nationally. Arizona's electric utilities are making efforts to attract "high load factor" customers, and generating capacity must be built to meet peak demands.

Access to the facility is an essential requirement. Many facilities have significant employment levels and operate multiple shifts, requiring a good system of surface transportation. Multilane streets, highway, and even freeway access may be required both for workers and for deliveries. Many companies want their facilities within a reasonable drive of a major airport. Appropriately developed sites and buildings, therefore, are necessary.

Infrastructure requires a significant capital investment as well as a commitment to fund ongoing operational and maintenance costs. Establishing a funding infrastructure in Arizona has not been easy. Arizona is severely limited in its ability to finance the development of public facilities, especially in advance of their need. Arizona's constitution places stringent limits on state debt, and subsequent limitations have been placed on municipal debt and spending. In rapidly growing areas, these limitations make it difficult to construct public facilities until after the population they serve is in place. The resulting inconvenience and increased cost is grudgingly accepted because the "newcomers" are paying their share of needed new facilities. However, in the rush to keep up with growth, adequate maintenance programs for public facilities have been overlooked. As Arizona's development matures, the state will meet the same maintenance, upgrading, and replacement issues faced by Eastern states decades ago, and Arizona will pay the price for deferring these expenditures.

RESEARCH AND DEVELOPMENT ENVIRONMENT

Dependent as they are on technological innovation, high technology companies choose to locate in areas where there is strong research and development activity. Much research is conducted by corporations, but the importance of a major research university is well known (Dempsey 1985; Rogers and Larsen 1984; Wigand 1985); over 70 percent of the respondents within the regional microelectronics industry indicated that the presence or nearness of Arizona State University ranked among the top three reasons to locate in the Phoenix area. University research activity can serve as a magnet to attract high technology firms due to the nature of the research and the presence of faculty and staff. University faculty members engaged in research sometimes may establish their own companies as a result of their experience. In other cases, industry may work cooperatively with university researchers, funding all or part of their work, supplying facilities and staff, or assisting in targeting the research to corporate needs.

Areas of high technology research and development specialization exist in Arizona, especially in the universities. Arizona State University (ASU) has developed the Engineering Excellence Research Center, which includes specialization in computer science, transportation, CAD/CAM, telecommunications, energy systems, solid state, and thermosciences. The University of Arizona has conducted research in optics, microcontamination, biomedicine, and astronomy. Northern Arizona University has specialized in the natural sciences and forestry.

Universities are also teaming up with developers, or becoming developers themselves, in undertaking projects to provide industrial or commercial space and incubator facilities. Some universities have established affiliates directly or by joint venture to conduct research and to provide specialized services to industry. These may have the effect of accelerating innovation while reducing the cost to companies of supporting the research program. It also creates revenues and develops properties adjacent to the universities. State government cooperates with Arizona's universities to develop projects that will induce desired high technology development.

ENGINEERING EXCELLENCE PROGRAM
AT ARIZONA STATE UNIVERSITY

In 1979 a decision was made to review the status of the College of Engineering and Applied Sciences, partially in response to the growing high technology industrial base forming in the Phoenix area. A fifty-member advisory council of engineering was organized, composed of leaders from high technology industry in Arizona, representatives of state government, and ASU engineering faculty. The initial goals of this council were to evaluate the engineering program at ASU and to develop a strategy bringing the College of Engineering and Applied Sciences up to national standards. The eventual goal was to make the College of Engineering and Applied Sciences at ASU one of the top schools in the country in graduate studies and a top research institution. The advisory council plan included several five-year phases.

Phase I

The Phase I plan called for total expenditures of $32 million coming from government and private industry. This goal was far exceeded with $54 million ($12 million in equipment) raised for the Engineering Excellence Program. (Phase II, described later, covers the period from 1985 to 1990.)

The structure of the College of Engineering and Applied Sciences has changed with the advent of the Engineering Excellence Program. To support the research component of the Engineering Excellence Program, four research centers have been developed: the Center for Advanced Research in Transportation, the Center for Solid State Electronics, the Center for Energy Systems, and the Center for Automated Engineering and Robotics. A fifth center in telecommunications is proposed. Together these academic and research units emphasized six content areas in Phase I: solid state electronics, computers/computer science, computer-aided processes, energy, thermoscience, and transportation.

The major strides accomplished by the College of Engineering and Applied Sciences in a short five-year time span are exemplified by the growth of the faculty, students, and physical plant. There are sixty-five new faculty lines (a 59 percent increase) and fifty-two (FTE) new graduate assistant lines (a 334 percent increase) (Beakley et al. 1985). Sponsored research has increased from just over $1 million in 1979 to approximately $9.5 million in 1984 (an 864 percent increase). The undergraduate population has increased from 2,547 in 1979 to 3,351 in 1984 (a 32 percent increase), and graduate enrollments have increased from 712 students in 1979 to 977 in 1984 (a 37 percent increase). Finally, the Engineering Research Center, which includes a 4,000 square foot class 100 clean room, portions of which are class 10.

The growing bond between the restructured engineering program at ASU and industry is seen in the development and implementation of a strong continuing education component. In 1980 a Center for Professional Development was established as part of the Engineering Excellence Program. The goal of this center is to meet the increasing demand by engineering and applied science professionals for continuous updating and maintenance of their technical competency and skill. From its inception to the end of Phase I in 1984, over 160 short courses and institutes were held and attended by more than 5,000 professionals.

The Interactive Instructional Television Program (IITP) began broadcasting courses to off-university sites in 1982. Computer science and engineering courses are directed to the high technology companies located in the Phoenix area. As of 1986 there were fifteen participating sites. Through a sophisticated teleconferencing system, students at the remote sites are able to interact with the faculty person giving the lecture and with students at other remote sites. This method of providing graduate-level courses and special seminars was initiated at the request of local industry so that their employees could receive graduate degrees.

Phase II

The accomplishments of Phase I have been considerable and the goals of Phase II (1985 through 1990) are equally ambitious. The efforts of Phase I placed the College of Engineering and Applied Sciences in the top twenty engineering programs in the country, according to a 1983 National Academy of Sciences study. Phase II seeks to obtain another $62.5 million for the Engineering Excellence Program to complement the $54 million already invested. This money is targeted for the construction of a $16 million Engineering Design Center and to expand the content areas from six to eight by including telecommunications and instruction. Government sources are expected to provide $22.5 million; the university will provide $20 million; and industry is expected to provide $20 million. By the end of 1984, $12.3 million in cash and equipment had been raised for the program.

The continued support of the Engineering Excellence Program comes at a time when state officials are predicting that approximately 90,000 individuals are currently employed in high technology jobs, and employment is expected to grow by 41 percent in the next three years. Phase II is designed to keep pace with this rapid growth, which demands roughly 2,000 new engineers annually (Winton 1985).

ASU AND THE MCC EXPERIENCE

ASU's engineering program played a central role in Arizona's effort to entice the Microelectronics and Computer Technology Corporation (MCC) to locate in Arizona, specifically the Phoenix/Tempe area. The MCC is a joint research and development venture created to help maintain U.S. technological preeminence and international competitiveness in microelectronics and computers. MCC's objective is to develop technology and development tools that will allow its shareholders to derive products and services of their own conception and design and to compete in markets of their choice.

MCC was chartered as a Delaware close corporation in 1982 and is located in Austin, Texas. MCC opened its Austin facilities in October 1983 and began research in January 1984 under the direction of retired Admiral B.R. Inman. MCC is owned by twenty-one U.S. companies, referred to as *shareholders*.

Arizona—especially the Governor's Office and the Office for Economic Development and Planning (OEPAD), now the Arizona Department of Commerce—had engaged previously in large industrial recruitment efforts. The MCC effort, however, has been described as Arizona's superbowl effort of recruitment. Reportedly, Governor Bruce Babbitt first learned about MCC and its search for a headquarters through an article in the *Wall Street Journal* during an industrial recruiting trip to San Jose, California, in January 1983. According to Larry Landry (1985), the former director of OEPAD, it was on this recruiting trip that the governor and business leaders decided to make a serious bid to attract MCC to Arizona.

A delegation consisting of Governor Babbitt and representatives of the Phoenix and Tucson business and academic communities flew to San Francisco less than two months later and made a presentation to the MCC. At that time, Arizona's effort was the only one in which a governor had led a delegation to the MCC. Arizona stressed its benefits as a desirable place to live (sunshine, recreation) and a good place to do business (strong labor force, good educational institutions, low taxes) but offered little in terms of tax breaks, free land, office space, and faculty endowments (Landry 1985). In comparison with bids made by other states, Arizona's bid was not especially lavish and was viewed by some as an accurate reflection of the legislators' mindset—that is, that "Arizona sells itself."

Admiral Inman came to Phoenix on March 17, 1983, for a speaking engagement and expressed that he was positive about Arizona's chances. One week later, however, the MCC announced the final cities in the competition and Phoenix was not one of them. Admiral Inman explained that Arizona was deleted from this list of finalists for two reasons: (1) Arizona did not have the educational excellence needed in the engineering and research areas, and (2) the state did not have the cultural resources and sophistication sufficient to attract the type of professional people that the consortium needed.

The reaction to this rejection was one of shock and constituted a rude awakening for many. The civic pride of business and community leaders was wounded. Although some tried to imagine what sort of incentive package would have attracted MCC (such as tax breaks, land and building incentives), much discussion focused on the role that quality of life and educational excellence play in economic development. Community and business leaders agreed that if Arizona wanted to be known as a high technology state, then a better understanding of what the industry expects in terms of educational excellence, quality of life, and cultural amenities was needed.

Consequently, two months after the MCC decision, a conference was held to explore issues affecting Arizona's high technology future and to develop a plan to address those issues (Landry 1985). This Governor's Symposium on High Technology was attended by 150 business, finance, education, and government leaders. Among the guests were W.C. Norris (Control Data Corporation), Belden Daniels (Council for Community Development), Gary Tooker (Motorola), and Dean C.R. Haden, ASU College of Engineering and Applied Sciences. These individuals discussed the role of education, manpower training, venture capital, and university/industry research partnerships in attracting and nourishing high technology industries. Subsequent small symposia and work groups developed specific recommendations in each of these areas. After specific recommendations had been made, a follow-up conference was held to further develop a "road map for Arizona's high-technology future" (Landry 1985).

The MCC decision jolted Arizona into acknowledging the fiercely competitive environment that exists among all fifty states in trying to attract high technology industry. This realization moved Arizona into a new economic development policy era. At the same time, business and community leaders used the MCC experience as an argument in developing new initiatives to enhance educational excellence and quality of life. Previously, most economic development efforts were essentially prospecting trips to attract industry or lobbying the state legislature for an increased state advertising budget. Economic development efforts now focused on upgrading graduate programs in the state's universities and enhancing the quality of education in general.

It should be noted that by the time the MCC decision was made, funding for ASU's Engineering Excellence Program had been secured and legislation was passed creating the ASU Research Park. In retrospect it seems that without these efforts Arizona probably would not have been in the initial competition for MCC (Landry 1985).

THE ARIZONA STATE UNIVERSITY RESEARCH PARK

Consistent with the emphasis on high technology research, ASU developed a research park to further increase the cooperation between industry and the university. The concept of research parks associated with universities is not a new one. The first, and possibly the best known, was developed by Stanford University in 1951 and continues as a prominent and successful park serving the Silicon Valley area. Today, there are over 300 research parks in the world, over half of which are located in the United States (General Accounting Office 1983; Owens 1985; Showalter 1986). Ideally, a park should bring university researchers together with their counterparts in industry, integrating the results produced by both parties. This approach encourages more university research and places the university on the cutting edge of new technological developments.

The development of the concept for the ASU Research Park was the product of a number of factors. The pressure placed on ASU by industry to improve the engineeering program was possibly the prime mover in the development of the research park. However, the actual proposal to create the park came in 1982 from the president of ASU and two businessmen. The feasibility of the park was studied by a committee composed of university and business representatives. In 1983 the ASU board of regents approved the proposed park. In 1984 the board of regents approved the master plan for the park and authorized the creation of a separate seven-member, nonprofit corporation, Price-Elliot Research Park, Inc. This corporation has as its mission the design, development, marketing, and administration of the park on behalf of ASU for ninety-nine years.

On December 4, 1984, ground was broken on 323 acres of land located in the southeast section of metropolitan Phoenix. Sky Harbor International Airport is approximately fifteen minutes away by car and ASU is five miles away. ASU and the nonprofit corporation invested $40 million in land and intrastructure for the park. Individual building sites on 206 acres, and several multitenant buildings with a total of 136,000 square feet of space are available to industry. Ultimately, the master plan calls for the development of 3 million square feet of space. The city of Tempe, in which the park is located, is improving the feeder roads surrounding the site.

The ASU Research Park offers tennis courts, a FAA-approved heliport, jogging paths, equestrian trails, ramadas and picnic areas, and three lakes that are fed by treated waste water from a nearby Motorola manufacturing plant. By the end of 1987 a conference center will be constructed, and a hotel site including a child care learning center, a health management facility, pool, restaurants, and support services are planned.

The Arizona University Research Park is still very young and has yet to mature; however, the barely two-year-old park is, by comparison, ahead of other U.S. research parks. For example, it took Research Triangle Park seven years to get its first tenant, and there were ten years between Stanford Research Park's first and second building. The next five years will reveal more conclusively the outcome of this university-originated, innovative venture.

CONCLUSIONS

In Arizona, university research centers have much to offer high technology industry, and the universities need the additional support that industry can provide. ASU's Engineering Excellence Program, now in its second phase, represents a solid beginning, but more can be done. Policies of Arizona's universities and board of regents need to encourage partnership approaches, patent filing, and research incentives. Barriers to cooperative efforts should be removed to the extent possible. Unfortunately, most university partnership efforts seem aimed at larger high technology firms with significant financial resources. Smaller firms also have much to offer, especially in the area of innovation, but they typically choose not to participate in university programs. Turf problems and interuniversity competitiveness also limit partnership arrangements. University activity in high technology seems to be limited to the community in which the university is located. Efforts to work with industry in cities where other Arizona universities are located are viewed as poaching. This parochialism limits the impact of university research programs and reduces opportunities for cooperative programs.

ASU's Research Park is a good step toward developing specialized facilities for industry with specific needs. Further opportunities should be explored to develop facilities for specific high technology industry in areas where a university has a research specialty. It also may be desirable for the state development office to work cooperatively with universities to develop facilities.

High technology industry in Arizona stresses manufacturing as opposed to research and development, and some have become concerned about the so-called branch town syndrome. Most private- and public-sector officials seem to think that this trend is likely to continue unless Arizona becomes more competitive in attracting the national headquarters of high technology companies with the corporatewide research and development activities.

More and more frequently industry spokespersons and public officials alike are worried that Arizona exhibits a laissez-faire approach to high technology development. They are concerned that Arizona must take a more active role in economic development, stressing the state's accomplishments and describing its economic realities. It has become obvious that the days of just sitting back and figuring that high technology firms will keep coming to Arizona are over. Sunbelt charm alone is no longer adequate in such a highly competitive environment.

Arizona is only one participant in the worldwide competition for jobs, factories, and services. A more active position is needed—one that stresses opportunities that exist in Arizona. Chief among these opportunities is the potential for conducting research, an activity that is knowledge intensive rather than resource intensive. The research university can play an important role in this process, and increased technology transfer between the university and industry and government may be a prime area of increased attention.

BIBLIOGRAPHY

Arizona Department of Economic Security. 1984. "Arizona Education—A Statistical Overview." *Arizona Labor Market Newsletter* 8 (June): 21–27.

Arizona Office of Economic Planning and Development. 1983. "Arizona Horizons: A Strategy for Future Economic Growth." Phoenix: OEPAD.

———. 1984. "High Technology in Arizona: A Market Analysis of Suppliers in Arizona and the Southwest." Phoenix: OEPAD.

Beakley, G.C., C.E. Backus, and R.W. Kelly. 1985. "Results of the First 5-year Phase: Excellence in Engineering for the '80s." Report prepared by the Dean's Advisory Council. Tempe: College of Engineering and Applied Sciences, Arizona State University.

Cole, J. 1987. "Median New Home Value Rises." *Arizona Republic*, February 15, pp. C1–C2.

Dempsey, K. 1985. "Hi-Tech Race on for Silicon Areas in U.S." *Plants, Sites & Parks* 12 (March/April): 1, 3, 17, 20–22.

General Accounting Office. 1983. *The Federal Role in Fostering University-Industry Cooperation.* PAD-83-22. Washington, D.C.: GAO.

Hall, A. 1985. "Americans Heading to South, Not West." *Arizona Republic* 23 (January): C1.

"Inside Phoenix." 1985. *Arizona Republic/Phoenix Gazette*, pp. 40–45.

Landry, L. 1985. Personal interview. November 17.

Larsen, J.K., R.T. Wigand, and E.M. Rogers. 1987. "Industry-University Technology Transfer in Microelectronics." Report submitted to the National Science Foundation, January.

Leinberger, C.B. 1984. "Urban Villages: The Locational Lessons." *Wall Street Journal*, November 13.

Lindsey, R. 1987. "Alarm Raised on Growth of Phoenix." *New York Times*, March 12, p. 15.

McNulty, M., and G. Woodard. 1984. "Arizona Water Issues: Contrasting Economic and Legal Perspectives." *Arizona Review*: 1–13.

Mountain West Research–Southwest, Inc. 1984. "Phoenix Metropolitan Housing Study, Fourth Quarter." Phoenix: Mountain West Research–Southwest, Inc.

Noyes, F. 1987. "Economic Growth Dependent on Quality of Life, Officials Told." *Arizona Republic*, March 31, pp. B4, B16.

Owens, R.W. 1985. Personal interview, October 21.

Price, K., and C. Hodge. 1984. "High-Tech: Valley Spinning Shimmering Web of Pure Silicon." *Arizona Republic*, July 23, p. D1.

Rogers, E.M., and J.K. Larsen. 1984. *Silicon Valley Fever*. New York: Basic Books.

Rogerson, P.A., and D.A. Plane. 1985. "Monitoring Migration Trends." *American Demographics* 7 (February): 27–29, 47.

Ronan, B. 1986. *A Summary of Arizona's Changing Economy: Trends and Prospects*. Phoenix: Commerce Press, Arizona Department of Commerce.

Showalter, J. 1986. "Officials Discuss Industry-University Marriage." *Tempe Daily News*, April 30, p. C8.

Stanton, S. 1984. "State's High-Tech Growth Requires Strong Pollution Rules, Panel Told." *Arizona Republic*, September 1, p. B1.

Steiner, W.E. 1985. *Water for Municipal and Industrial Growth*. Phoenix.

University of Arizona College of Business and Public Administration. 1984. *Arizona Economic Indicators* (Spring/Summer): 1.

Valley National Bank. 1986. *Arizona Progress* 41(9): 2.

———. 1984. *Arizona Statistical Review* (September): 20.

Vest, M.J. 1985. "The University of Arizona's Economic Outlook 85/86." Unpublished report, University of Arizona, February, p. 1.

Webster, G. 1987. "En-tribe-preneurs: Hopis Use Ancient Skills to Master High-Tech Jobs." *Arizona Republic*, March 1, p. E1.

Wigand, R.T. 1985. "High-Technology Development in Arizona." Working Paper, Arizona State University, Phoenix, August.

Winton, B. 1985. "ASU to Pursue Phase Two of 10-Year, High-Tech Plan." *Tempe Daily News*, March 6, p. 1.

Part III

TECHNOPOLIS:
ISSUES AND CONCERNS

Chapter 12

BASIC SCIENCE AND
THE TECHNOPOLIS

Hans Mark

The objective of this chapter is to emphasize the link between the resources of universities and the development of technopoleis, which to me is really no different from that of cities. Civilization depends on the existence of cities, and cities have always been associated with the development and then the exploitation of technology. The concept of *technopolis* is an ageless topic. I lived through the Route 128 phenomenon when I was a graduate student at MIT in the early 1950s. Then I moved to California in time to watch Silicon Valley grow. People who live through a period of history usually begin to understand, in depth, that period only after it has passed.

THE REAL RESOURCES OF A UNIVERSITY

It is important to understand what the real resources of a university are. Universities look impressive because they have a lot of buildings and football stadiums. But the real resources of a university are a few very bright people on a faculty. There are not very many of them. A major university might have twenty or twenty-five people on its faculty of about 1,000 who by any standard could be called real resources. The Japanese have a nice way of recognizing this; they actually designate certain artists and certain people's accomplishments as national resources. There are not very many of them; they are very unusual people.

What some of these unusual people at The University of Texas at Austin are doing right now has little to do with what business and community leaders may be interested in today—which is how to organize themselves to make money. But it has everything to do with how we are going to make money fifty years from now. I would like to make some predictions and provide some examples of some folks who work and live in the Austin/San Antonio Corridor and who, in their own minds, already live in that future and are doing something about it today. It is a justification for basic research, which is a major function of the university.

BASIC RESEARCH AND ECONOMIC DEVELOPMENT

Recently a supernova—a star that flashes and emits vast amounts of radiation for a while and then decays—was observed. At The Univeristy of Texas at Austin there are people who are thinking about supernovi because one of the remnants of some of these supernovi is a black hole. In a black hole the density of matter is so great that light, as we know it, cannot get out. As light tries to get out, it is attracted by the gravitational force of this black hole so strongly that the photon cannot leave and, consequently, cannot be seen. Why is it important to look at something that cannot be seen? In this black hole there is a state of matter that is entirely different from anything that we are familiar with in ordinary life. It is a state of matter where a cubic centimeter has the approximate weight of our sun. If you look at this matter in equilibrium—that is, look at it as a gas—then the particles that make up this matter have nothing to do with electrons, protons, neutrons, and the kinds of things that we are just barely beginning to understand.

Right now, almost our entire knowledge of this phenomenon comes from astronomy, or that most impractical of all sciences. I predict that fifty years from now somebody is going to make money from the knowledge that comes from black holes. Why am I so confident in making that prediction? The first industrial revolution was based on mechanics, and if you look at the people who codified mechanics before 1750, they were people like Johannes Kepler and Isaac Newton. Their codification of mechanics was based almost entirely on astronomical measurements. The reason that I think that I am fairly safe in predicting that somebody is going to make money out of black holes is that it has happened before. People made money out of Newtonian mechanics and differential calculus, something derived in the late years of the seventeenth century through astronomical measurements and knowledge alone.

At The University of Texas at Austin people like Nobel Laureate in Physics Steven Weinberg and Distinguished Professor in Physics John Wheeler are thinking about these problems. While we go about the business of creating the Route 128s and the Austin/San Antonio Corridors, it is important that we take time to listen to such folks as these and find out what they are saying. Their research is where, at some point in the not-too-distant future, the next fundamental developments will come from. I have mentioned two people at The University of Texas in Austin. Because one of the chapters in this text discusses the Austin/San Antonio Corridor as an emerging technopolis, let me talk about somebody at The University of Texas in San Antonio and also make some speculations about what might come of his work.

Today's computer is basically an array of switching elements. A microchip will have something like 10^{13} or 10^{14} atoms in it; it is pretty small. The performance of a computer depends on the size of the switching element. The more switching elements packed into a given volume, the more powerful the computer. I asked my colleagues who know something about quantum mechanics and about solid state physics, "How small can you make a switch?" They tell me that we might be able to go down two orders of magnitude to maybe 10^{11} atoms. After that, you get into trouble because when these arrays of atoms become too small, the regularities on which the performance of the switch depends disappear. Edge effects begin to be a barrier. From the point of view of making an investment that will pay off in five years, if you know there are still two orders of magnitude of progress on the technical horizon, that is a pretty good reason for making an investment in computers—to try to make more powerful and smaller computers.

However, if we are interested in the longer term, we might refer to the work of a faculty member at The University of Texas at San Antonio, Professor Matthew Wayner. He is both an electrical engineer and a biochemist and is researching the problem of how electric currents are controlled and switched in biological systems. It has long been known that the switching elements that control the currents that run inside our nerve fibers have something of the order of 10,000 atoms and not 10^{13}. They are, in short, *nine orders* of magnitude *smaller* than the switching elements in our best computers today. They are different: They do not control electronic currents; they control ionic currents. We do not know yet how to work them or switch them. Professor Wayner is isolating these molecules and trying to impregnate them on membranes of some kind and then learning how to control them so that it is possible to build characteristic curves that are common to any nonlinear electronic element. If this faculty member succeeds, the important application that is obvious is the one for making smaller computers, which is fine. However, that is not the real thing that is going to come out of this development.

Two of my colleagues at UT Austin—one in electrical engineering and another in computer science—and I began to speculate about what happens when you have really small switching elements. What happens is comparable to the human brain—the brain probably has somewhere between 10^{11} and 10^{12} of these little molecules that switch currents. Currently a computer has several million switching elements. A computer modeled after the human brain would have six or more orders of magnitude of more switching elements. We may speculate about whether this quantitative difference has qualitative effects in how that machine can perform. Can it be hooked up in ways that are different? Is there an organizational principle that we do not understand yet that has to do with the qualities that we call judgment or imagination? Does simply having more switching elements hooked up in some random fashion lead to these qualities? If it is so, that is another one of the insights that come from a university, where clearly somebody is going to make some money fifty years from now.

CONCLUSION

Universities are crucial not only in terms of the trained people that they turn out who are necessary to support a technopolis, but in terms of how we sustain the capital that the university provides, Walt Rostow's (UT-Austin professor of political economy) capital—namely, the new knowledge. It is important to know about black holes, not only because it is scientifically important but because of practical applications in the future. For this same reason it is important to understand how molecules switch currents because of the basic scientific question of how the human mind works and the practical application of how that knowledge is turned into new computing machines. In short, all economic development ultimately comes from human imagination. It is the development and the exploitation of that all-important organ, the human mind, that is so tremendously important in the success of technopoleis.

Chapter 13

IMPLEMENTING A HIGH TECH CENTER STRATEGY
The Marketing Program

John K. Ryans, Jr., and William L. Shanklin

Competition for economic development by nations, states, and countless communities is more intense than it ever has been as these entities compete for industry and jobs and the tax revenues that result. A centerpiece of a growing number of these economic development efforts and programs is a tech center or centers. Articles in such influential periodicals as *Fortune* (March 16, 1987), *Insight* (December 22, 1986), *Industry Week* (September 15, 1986), and *Business Week* (January 13, 1986) are indicative of the attention that tech centers are receiving. The *Fortune* article reported that, as of early 1987, there were eighty university research parks, up from only a dozen in 1983. And this eighty is not inclusive of tech centers that are not university related or are not research parks.

The tech center concept is hardly revolutionary. What is new, however, is the large and increasing number of private- and public-sector entities that intend to develop, or have already done so, tech centers. Indeed, the tech center strategy for economic development is no longer unique or confined to a few geographical locales.

In our view, there are key market- and marketing-related questions that, if addressed in the exploratory stage of development, will mitigate the considerable risks of implementing a tech center. It is imprudent and potentially costly for a sponsor (public or private sector) to hastily decide to go forward with a tech center without initially investing in some objective marketing research concerning a center's viability.

Prior to committing too much in the way of effort and dollars, a tech center sponsor would at least want to obtain the best objective information feasible pertaining to several issues on which success or failure can depend. For instance:

1. Is there a demonstrable need or demand for a tech center in the geographical area of concern?
2. If so, what kind of tech center is needed in terms of technology—all encompassing or focused?
3. Does the state and community have the right ingredients (such as cash, engineers, scientists, a skilled work force) necessary to make a center effective?
4. What are the marketing considerations that pertain to tech center operations?

These are the types of issues that this chapter examines. The most rudimentary and probably the most complex issue of all, however, is the kind of tech center that the sponsor intends to have.

VARIETIES OF TECH CENTERS

The phrase *tech center* can be applied to many different types of high tech initiatives by the public and private sectors. A tech center can mean a research facility at a university, a science-oriented industrial park, or numerous other variations.

A valuable explanation of types of high tech initiatives by states, communities, and the private sector is provided by a 1984 report of the U.S. Office of Technology Assessment (OTA) (1984) entitled *Technology, Innovation, and Regional Development: Encouraging High-Technology Development.* Briefly, here are some of the important distinctions that the OTA report draws.

High technology initiatives in the United States have been sponsored by the states, local governments, the private sector, and combinations of these. Federal funding of R&D is also a high tech initiative, and federal research facilities like those run by NASA can be viewed as giant tech centers. However, the discussion here is focused on tech center efforts by state and local governments and the private sector. We have focused our recent research on the university-center concept and will use the insights we have gained for much of our later discussion.

State high technology initiatives are, of course, by no means confined to tech centers. For instance, states are involved in programs that run the gamut from capital provision and assistance programs, which attempt to assist high tech entrepreneurs in finding venture capital, to providing technical expertise to these entrepreneurs. Whenever a state government becomes involved with tech center programs, it is normally in conjunction with the state's public and private universities.

These state/university joint efforts can take several forms. We want to mention some of the major types of endeavors.

Research and science parks are clusters of research-intensive companies and facilities on a site near a university or universities. This arrangement is meant to foster transfer of information and knowledge between education and business. The Stanford Research Park in California is the oldest university-based science park, but similar efforts are underway across the United States, recently, for example, at Virginia's Center for Innovative Technology near Washington, D.C. An important byproduct of these kinds of centers is intended to be new businesses based on high tech discoveries made in the university and corporate labs.

Another form of university-based tech centers is called *research and technical centers* by the OTA. These conduct applied research for fees and other support. Examples are the Microelectronics Center of North Carolina, a similar center at the University of Wisconsin, and California's MICRO Research Center. In Ohio, the state's Thomas Alva Edison Partnership Program created six advanced technology application centers that are run by seven Ohio public and private universities. Each center specializes in research and development in a particular high tech area of expertise. For example, the center in Akron-Cleveland is concerned with polymer R&D and is widely referred to as an attempt to create a polymer valley.

A number of universities have implemented incubator facilities, which are at least quasi-tech centers. These incubators provide services to new high technology businesses or to aspiring entrepreneurs, ranging from low-cost office space to technical and management counseling. The oldest such facility is the University City Science Center in Philadelphia, founded in 1967 by twenty-three colleges and universities.

Local governments also have been active in fostering tech centers. In a few cases they work virtually by themselves but far more often in conjunction with state government, and there is almost always involvement of a local university. Local governments can support tech centers through assisting monetarily with university improvements, such as in improving an engineering school, supporting vocational training, helping with market research intended to determine the feasibility of a tech center with marketing programs meant to lure high tech firms to a tech center facility, initiating and shepherding high tech industrial incubators, and facilitating land use, planning, and zoning, with careful concern for high tech firms' requirements.

Looking at the experiences of tech centers across the United States, it is our view that some degree of participation by the private sector is vital. At the state-initiated Advanced Technology Development Center located at the Georgia Institute of Technology, the private sector is contributing heavily in terms of dollars. And perhaps the best case-in-point is the Research Triangle in North Carolina where there is public-sector/private-sector cooperation. Two state universities, North Carolina and North Carolina State, participate with a private university, Duke, and the firms that are located in the Research Triangle Complex. Similar successes at Rensselaer Polytechnic Institute, the University of Pittsburgh and Carnegie-Mellon University, and the University of Central Florida further emphasize the value of such linkages.

According to the OTA report (1984), the local communities that have benefited the most from private-sector high technology initiatives have three commonalities that appear to underlie success:

1. An organizational culture that promotes a common civic perspective and a positive attitude about the attributes and prospects of the region;
2. An environment that nurtures leaders, both public and private, who combine an established track record for innovation and entrepreneurship with a broader view of their community's resources and promise;
3. A network of business and civic advocacy organizations that attracts the membership of top officers of major companies and receives from them the commitment to work on efforts of mutual concern, including cooperation with the public sector.

Essential Questions to Ask

The several varieties of tech centers to choose from and the several combinations of state, local, and public-sector/private-sector sponsorships possible underline why precision and specificity are needed by anyone who is thinking about starting a tech center. Various kinds of tech center commitments— for example, a research and science park versus a university-based tech center—require considerably different monetary outlays and far different tasks performed by the sponsoring organization(s).

We know that realistically addressing a number of questions early on in the planning phase for a tech center can help to prevent a host of unfulfilled expectations later.

1. What are our strategic competencies? In other words, what can our state, our community, or our organization offer in the way of engineering and scientific expertise, a skilled and educated work force, venture capital, and managerial talent?

What makes us especially qualified to launch a tech center? What is our competitive advantage over other areas of the country that might also start one?

Several states, local communities, and universities have started research parks, for instance, with little chance for success because the critical ingredients for making a tech center work are absent. Certain kinds of technologies require the presence of a "global caliber" engineering or science faculty at a university or facility such as NASA, a pool of highly skilled technical workers, a supporting high-quality primary and secondary educational system, and plentiful venture capital. The Route 128 area in Boston is successful partially because of MIT, Harvard, and other area universities, as well as the right mix of ancillary factors that combine to make Boston a high tech center, including being a place where it is easy to recruit technical professionals because of the amenities of Boston. Few other parts of the United States can ever hope to duplicate the Boston success story. Even though many are trying, such as Mississippi's Institute for Technological Development,[1] most will not have all that much to show for their time, effort, and dollars.

Most states, local governments, and private-sector efforts would be better off if they were more realistic on the question of strategic competency and therefore did not limit their development goals to high technology. In fact, evidence mounts that the job-generating capacity is much greater for medium to low technology endeavors than it is for high technology.

2. Assessing these strategic competencies, is a tech center feasible at all? If so, which kind or kinds?

Based on hard evidence and a realistic assessment, if a potential sponsor concludes that a tech center of some kind is workable (and most should not), then should the tech center be a research institute at a university or should it be something more ambitious on the order of a heavily funded science park? Here, the issue is the degree of monetary outlay in bricks and mortar, administration, capital requirement, personnel, marketing, and market research needed to make the tech center effective in reaching the goals set for it.

Just as the aging starlet and the high school football hero may have a difficult time seeing reality in their mirrors, communities may also have difficulties in separating dreams and aspirations from the real world of fostering high technology economic development. We continually see and hear of business leaders and public officials deluding themselves into thinking that their local university, college, or even community college is a plus in attracting high tech companies or in providing a home for first-level research-oriented technical educators.

Simply stated, there is no substitute for (1) a thorough, objective local resource market audit and (2) a comprehensive competitor analysis, prior to any "go ahead" decision for a tech center. It is important to recognize that marketing's role begins at this point rather than in the later stages when the plans for the tech center are in place. When we conducted a 1987 survey among university-related technical centers regarding their marketing activities and marketing problems, an all too frequent comment we received went something like, "We have not yet developed to the stage where we would need to be concerned about marketing." In brief, marketing was being thought of as, and consequently limited to, promotional activities; the value of market and marketing research in identifying opportunities and threats before a decision to launch a tech center was often forgone.

3. Should the tech center focus on a general high tech approach or be limited to one technology?

A number of states and communities, as well as private organizations, have been directed in their high tech development initiatives; others seem to prefer what might be termed the "casting a wide net" approach. As Table 13–1 shows, states tend to target more than one high tech industry. The larger, more diversified states like California or New York, have a broader targeting approach that appears consistent with the heterogeneity and immensity of their resources, interests, and university-related strengths. Similar broad targeting tends to make much less sense, however, on the regional or community level.

We have examined high technology from several viewpoints. While conducting research for our two books, *Essentials of Marketing High Technology* (Shanklin and Ryans 1987) and *Guide to Marketing for Economic Development* (Ryans and Shanklin 1986) we considered how high tech industries develop. We feel that a single high tech industry strategy (such as Akron's Polymer Center and Irving, California's Bioscience Center) offers far more benefits to most communities and states than does a more generic high tech approach. We see the benefits to include

1. The ability to continually attract the prerequisite academics and educational resources;

Table 13-1. Targeted High Technology Industries and Business Activities in Selected Survey States.

	CA	CT	GA	MA	NY	NC	PA[a]	TN	IL	IN	MI	MN	MO	NM	OH	RI	Total
Targeted high technology industries																	
Space/avionics	X	—	X	—	—	—	—	—	—	—	—	—	X	—	—	—	3
Transportation	—	—	—	—	—	—	—	—	—	—	—	—	—	X	—	—	1
Communications	X	—	X	—	—	X	—	X	—	—	X	—	—	—	—	—	5
Electronics	X	X	—	—	—	—	—	—	—	X	X	—	X	—	—	—	5
Microelectronics	—	X	—	X	—	X	—	—	X	—	—	—	—	—	—	—	4
Robotics	—	—	—	—	—	—	—	—	—	—	X	—	—	—	—	—	1
Computer hardware	X	—	X	X	X	X	—	—	X	—	—	—	—	—	—	—	6
Computer software	—	—	X	—	X	—	—	—	X	—	—	—	—	—	—	—	3
Lasers	—	—	X	—	—	—	—	—	—	—	—	—	—	X	—	—	2
Energy	—	—	X	—	—	—	—	X	—	—	—	—	—	—	—	—	2
Biotechnology	—	—	X	X	X	X	—	X	X	—	X	—	X	—	—	—	7
Biomedical	—	X	—	—	X	—	—	—	X	—	X	—	X	—	X	X	7
Pharmaceutical	—	X	—	—	X	X	—	—	X	—	—	—	—	—	X	—	4
None targeted	—	—	—	—	—	—	—	—	—	—	—	X	—	—	—	—	1
Targeted business activities																	
Manufacturing	X	X	—	X	—	X	—	X	—	X	X	—	X	X	X	—	10
R&D	X	X	—	X	—	X	—	X	—	—	X	—	X	X	—	—	8
Services	—	—	—	—	—	—	—	—	—	—	X	—	X	—	X	—	3

[a]Pennsylvania has targeted twenty-seven specific industries.

Source: Research Triangle Institute, reprinted by Office of Technology Assessment.

2. The ability to offer the type of reputation enhancement that high tech firms often seek from being part of a high tech industry concentration—such as "We are located on Route 128 or in Silicon Valley";

3. A synergy in state (and company or local) training programs that results in the presence of a trained, local nucleus in a state-of-the-art field; and

4. The opportunity to specialize in and consolidate infrastructure requirements—that is, the complementary firms would have similar power, water, and waste disposal, needs.

Thus, there are many significant points favoring a single or limited high tech industry approach for a tech center.

On the negative side, there is an overall risk of failure associated with this limited technological focus that may at least appear to be greater than the risks of a more generic diversified strategy. As a hedge, the tech center strategy could involve two or more closely related industries or a high tech plus medium to low tech combination.

4. How do we set priorities?

Naturally, a corollary question to the technology-specialization issue concerns those locales that actually may have the resources necessary to pursue several technologies. Although many states and communities are fortunate to identify even one technology in which they have the requisite criteria for success, some states and cities can and do explore several avenues. Selection of one or a limited number of technologies becomes critical in this situation and demands a carefully developed prioritization strategy. After local strengths are fully determined and a competition analysis made, then the focus must shift to the prospective industries themselves. The crucial question becomes, Which industries have the greatest potential for immediate success, and which will require a longer development period?

5. Have we integrally involved enough influential people from the public and private sectors?

It is essential from the very outset to develop a harmonious working relationship among all parties likely to be involved in and concerned about a tech center's development. Frequently, the efforts of an area or region are far too fragmented to achieve real success. Even though a tech center may be university-based and be funded by a state/private-sector consortium, it still needs the cooperation and support from the local community. Similarly a group of local business and government leaders cannot "charge ahead" and make tech center plans without the active involvement of the university community, which may in fact be the linchpin needed to make the effort workable.

TECH-CENTER MARKETING ISSUES

The extent to which a tech center achieves success in meeting its goals depends to a large measure on its effective use of marketing. Marketing research and environmental scanning initially provide the information needed about potential demand for a high tech center, as well as what kind of center might work best considering the state, community, and/or university's strategic competencies. Given this data, the interested parties can then develop realistic objectives for the center and proceed to formulate strategies to achieve the objectives.

Consider just two questions that we believe illustrate the unique perspective marketing offers to high tech–center planners and operators. Here, we draw on the findings from our current research investigation concerning how marketing is being used in tech-center development efforts.

1. What marketing research tools are appropriate for focusing center planners on the appropriate technology or technologies to be included in their particular center?

Savvy consumer and industrial marketers are aware of their internal and external strengths and weaknesses as they seek to compete for sales and profits. In addition, they continually assess their competition and the general marketing environment in which they operate. The use of evaluative techniques that have been tried and proved in the private sector, notably the marketing audit, are appropriate for the tech-center planning process. The various components of the total marketing audit typically consist of the

1. Marketing environment audit,
2. Marketing strategy audit,
3. Marketing organization audit,
4. Marketing systems audit,
5. Marketing productivity audit, and
6. Marketing function audit (Kotler, Gregor, and Rogers 1977: 25–44).

The first two components are critical as the center determines the factors that will influence its success: its external environment, competition, competencies, and strategies. Later, the marketing audit process can be useful in monitoring the extent to which progress is being made and sustained.

Similarly, the techniques typically employed in competitive analysis could assist the center in evaluating the progress of competing centers. A competitor analysis would also help to determine the evolutionary stage of prospective technologies, like the incipient biotechnology field, and high tech companies, as well.

2.   Are there substantive marketing advantages to employing a single or limited high technology strategy for a tech center?

There are a number of reasons why individuals charged with marketing a high tech center may find a single or limited technology strategy best. Many of these reasons relate directly to efficiency and effectiveness. Take one important example. Personnel involved in marketing are, of course, familiar with the importance of building awareness and image. Moreover, it is crucial to establish credibility when communicating with prospective firms. These promotional efforts tend to be greatly enhanced by a single or limited technology strategy. It is far easier to establish and portray a tech center's expertise in one technology than it is to develop an image that the center has expertise in multiple high tech fields.

Other marketing advantages to employing a single or limited technology strategy include

1.   Effective and less costly promotional efforts (for example, advertising can be placed in specialized professional/trade publications rather than in general publications such as *Area Development* and the *Wall Street Journal*);
2.   Creative use of benefits in promotion and qualitative marketing research in assessing a prospective firm's needs (for example, the high tech center can more tightly position its "offering" when data obtained in research are specific);
3.   Improved criteria for selection of advertising agency and even high tech center marketing personnel can be employed; the individuals and agency selected can be judged on industry-specific experience, thus enhancing the likelihood of product knowledge;
4.   Closer fit between the proposed financial package offered by the community and the real concerns and demands of the prospect firms can be achieved due to a better understanding of the prospect's situation;
5.   Permit the center's trade missions (sales force) to be better prepared when visiting prospects and similarly to be more knowledgeable and effective in

using such media as trade shows and video tapes of the center/community for prospecting.

CONCLUSION

A good share of tech centers are making rather limited use of their potential marketing arsenal. Many centers appear not to have used marketing research techniques in seeking to assess their current local and competitive situations or to identify an appropriate high technology industry or industries.

Furthermore, marketing's role in high tech centers should not end once a center is initiated. In particular, at not-for-profit research organizations (such as at universities), there is the tendency to forget one important thing. That is, states and communities are interested in tech centers largely because, if successful, they can create new industries and revitalize others that generate jobs. Thus, the commercialization or marketing of the output of the tech centers is of vital importance.

NOTES

1. For example, see the advertisement for the Institute for Technology Development (Jackson, Mississippi) in the January 1987 issue of *Area Development* at page 73.

BIBLIOGRAPHY

Kotler, Philip, William Gregor, and Williams Rogers. 1977. "The Marketing Audit Comes of Age." *Sloan Management Review* (Winter): 25–44.
Ryans, John K., Jr., and William L. Shanklin. 1986. *Guide to Marketing for Economic Development.* Columbus, Ohio: Publishing Horizons.
Shanklin, William L., and John K. Ryans, Jr. 1987. *Essentials of Marketing High Technology.* Lexington, Mass.: Lexington Books.
U.S. Office of Technology Assessment. 1984. *Technology, Innovation, and Regional Economic Development: Encouraging High-Technology Development.* Background paper 2, OTA-BP-STI-25. Washington, D.C.: U.S. Government Printing Office.

Chapter 14

INTELLECTUAL PROPERTY AND COOPERATIVE R&D VENTURES

William G. Ouchi

Every society has had the problem of creating new intellectual property while balancing two incompatible objectives. One objective is to find a way to grant monopoly rights over the fruits of invention to the pioneer. If rights cannot be granted, the pioneer will not pioneer. On the other hand, every society desires to find ways to withhold monopoly from the pioneer to make the fruits of that new invention most broadly available to the public. At every period in history we find a different structure for achieving this balance.

For example, in the premedieval period, the primary manufacturing industries were textiles and the tanning of leather hides. At that time, almost all enterprises were family owned. The R&D budget consisted of sending a twenty-two-year-old son or daughter off for six months to visit the neighboring villages to see what he or she could learn. The expectation was that they would come back with some unique knowledge about how to more quickly tan a hide, how to produce a more supple hide, or how to achieve a new kind of weave on a loom. That would result in a monopoly for the family business for some period of time (80 percent gross margins and happy days for a couple

This chapter updates a report presented to the board of directors of the Semiconductor Industry Association and to the board of directors of the Semiconductor Research Corporation (SRC) at the end of 1986. This work has been funded by a grant from the SRC, which is a collaboration of about forty of the major integrated circuit manufacturers in the United States that have come together for the purpose of funding research and graduate training programs at universities in those areas that they feel are most important in their industry.

of years). The method for the protection of that property right was the kinship system. There was a chance that the offspring could be disloyal and reveal the process to competing neighbors. But the punishment for doing so was reasonably severe and therefore was not a common occurrence. It was a system of balance. It worked.

In the late medieval period, the first English merchants set sail for the Far East. They gathered a hundred souls, bought a piece of capital known as a 200-foot sailing ship, got on board, and set out for the horizon with the expectation of being eaten by a sea serpent. Should they have the good fortune to avoid that fate, the expectation was that they go off the end of the world. These people were real risk takers. This was real entrepreneurship. This was not fooling around with someone else's money. A small number of these merchants made it all the way to China, loaded up with spices, brought their cargo home, and were instant millionaires—just like the Silicon Valley kings. Suddenly they enjoyed great wealth. After a period of high living, they would assemble a new crew, outfit a new ship, and once again leave their home harbor. However, when they looked back over their shoulder, they saw a flotilla of imitators ready to follow them along their trading route. They immediately went back to the king and said, "If you think that I am going to go out there and take all of this risk so that these free riders can copy my discovery while taking none of my risks and share my rewards and reduce my profits, you must be crazy." In 1331 the king of England issued the first letter of protection, the forerunner of today's patents. He issued it to a textile maker, granting him monopoly rights over his "technology" for a period of sixteen years.

Meanwhile, technology in the manufacturing industries had become more complex. The family in its natural descendancy could not yield sufficient sons and daughters with the aptitude and interest to learn the required specialized technologies. So the ownership of R&D passed inevitably from the family to the craft guild. Craft guilds had the incentive and means because (1) they could monopolize R&D benefits because (2) they had sole control over who was admitted into the guild, (3) they could control who was taught the intellectual property, and (4) they enjoyed sole placement rights over where those people went to work and the terms and conditions under which they were employed. Again, it was a system of balance.

During the middle and late medieval period and early into the Industrial Revolution, the equivalent of Silicon Valley venture capital firms were the banks of the river Clyde. The start-up companies were ship builders, all competing against one another for better ways to lay a keel, to design a faster and more stable hull that would hold more cargo, to develop new materials and cuts for sails, and so on. As it turned out, in order to build the ship that did the job, you had to integrate the technologies. Shipbuilders had to bring together craftsmen from four or five different guilds and have them work together to jointly produce an intellectual property. Now, it happened that the traditional structure of the guilds was such that they were incapable of working together closely enough. Therefore, for the first time, we had the creation of an R&D department that was a group of people within the firm whose sole task was to invent new, integrated, intellectual properties that were owned not by a guild but by the firm.

In retrospect and oversimplifying it, Karl Marx argued that the inevitable consequence of industrialization would be that the worker would be alienated from the fruits of his labor. Another way to interpret that same event is that what happened was that the craft guilds were alienated from their monopoly rights over intellectual properties. Those monopoly rights instead were vested in the owners of capital. That transfer of intellectual property rights precipitated an immense transfer of wealth away from working guilds to the owners of capital and brought about a restructuring of society. This is to make a point that the structure of intellectual property rights that resides in any society is a tremendously powerful force. One, therefore, does not take it lightly. On the other hand, one should expect that the change in the structure of property rights is a fundamental fact of social-economic evolution.

As a consequence of these several developments, we have well developed in Western common law, and in Spanish law as well, the idea of private property. We learned to recognize the fiction that a person has a natural right to a house, a car, or a patent. We understand that in reality it is fiction that we entertain because no individual truly possesses such an inalienable right. We observe frequently that governments can exercise eminent domain proceedings and confiscate property. Governments can tax income and remove the wealth of individuals. On the other hand, by social agreement we understand that a person takes better care of something that he owns than something that he rents. Therefore, society lets people pretend that they own companies because then they will take better care of them. We understand that set of incentives. We understand the goals. We created a whole body of law that says if you put your own capital at risk and come up with a new intellectual property, there is the expectation of deriving monopoly rights from it for seventeen years under patent protection. The system worked well.

Another big change occurred in 1862 when President Abraham Lincoln signed into law the Morrill Act. The act established land grant institutions or the giving of land to make state universities. It set aside some 14 million acres of federal lands to be sold, the proceeds to be invested in securities. The income from the securities was to be used to support the establishment of institutions of higher learning dedicated to agricultural or mechanical arts.

The first land grant was received by the legislature of the state of Massachusetts and was used to establish the University of Massachusetts. The second land grant established MIT, which was chartered in 1862 but did not open its doors until the end of the Civil War two years later. It was under the land grant influence that the United States had a quintupling of the number of engineers produced in this country over the succeeding three decades. It was through this mechanism that the United States recognized the principle that we should take public funds and create public property, which we called basic research. It was not subject to private property rights and not subject to patent protection. It was freely available to all with the understanding that members of society could draw on that property created by tax funds to create wealth for themselves. New hybrids of corn would more directly enrich a farmer in Iowa than a street sweeper in Brooklyn, but we nonetheless felt that the benefit was sufficiently widespread and important to justify the transfer of wealth. Furthermore, it was realized that the incentive to invent a public property that others could readily use was so weak that if left to private markets, these new properties would not be created at socially optimal levels.

Therefore, society recognized the value of the principle that we would use public tax funds to create new intellectual properties that would ultimately flow to enrich only some individuals. It was a monumental move for U.S. society to take. This still is not done in Europe, and it still is not done in Japan. In fact, it still is not done in most countries of the world, in part, because most countries of the world have, in the past, had times when their universities became heavily politicized and became partisans in revolutions and social upheaval. The populaces of these countries feel that the universities should be kept away from influence of the business community and labor. They therefore passed philanthropic laws that discouraged the giving of money for research from business to the universities. Yet many people would argue that a great part of the vitality that has driven the United States for the last 200 years has been precisely from that combination.

In fact, when we look at the pioneers of Silicon Valley, whom we often think of as having succeeded by their individual hard work, risk taking, entrepreneurship, and so on, I think rather of the metaphor of the 49ers during the California gold rush. They were entrepreneurs. They were individuals, risk takers, and they worked hard. I take nothing away from them. But they didn't put the gold into the mountain. When I see 500 entrepreneurs all succeeding at the same time in the same place in the same industry, I get suspicious. My question is, "Who put the intellectual gold in that mountain?" We collectively put it there. It is part of the Morrill Act, the universities, the National Science Foundation, and other kinds of efforts in which we have created an unrivaled source of intellectual property in this country.

HORIZONTAL AND VERTICAL INTEGRATION

Let me tell you my version of the problem in the U.S. semiconductor industry and why it is so difficult to contend with. My recent research has focused on chip vendors. They build their chips with the tools that they buy from more than 500 U.S. vendors who have average annual sales of $11 million and are widely thought to be falling behind the Japanese in technology. Nevertheless, the U.S. chip maker still says,

> I really don't want to buy my next generation tools from an overseas vendor. The reason is not nationalism. The reason is that the one thing that I know about a new generation tool is that it never works quite right because it represents a new chemistry and a new process. Furthermore, I typically withhold information from my vendor when I tell him what I need so that he won't be able to tell my competitors what I am planning to produce. Because of this the vendor can't build the tools that I really need because I don't tell him what I really need. For all of these reasons, I don't get what I really need.
>
> If the chip building machinery fails and if my vendor is twenty-five miles away, I pick up the phone and say, "It doesn't work." He puts three guys in a car and they are in my plant that afternoon to make it go. If my vendor is 8,000 miles away and doesn't speak my language and my competitor is next door to him, then I am in deep trouble. Therefore, I really don't want to buy my next generation tool from an overseas vendor.
>
> On the other hand, the Japanese vendor has a better tool today than the U.S. vendor. If I don't buy it and my competitor does, I lose today. As much as I hate to say it, I have no choice but to form a strategic alliance with my key overseas vendors in order to protect myself, even though I know that in so doing, in the long run I shoot myself in the foot.

Also, during my interviews I hear the systems companies—the people who build computers—saying,

> I really don't want to buy my next generation microprocessor from an overseas vendor because I know that it is not going to be designed quite right the first time around. If my vendor is five miles away, I can pick up the phone and yell at him and he will send his guys over that afternoon, etc. On the other hand, the overseas vendor has a better device today. If I don't buy it and my competitor does, I lose today. Much as I hate to say it, I am going to have to form a strategic relationship with my key overseas vendor and let the U.S. industry go, even though I know in so doing, I shoot myself in the foot.

It is just a matter of time before the end users of the world start to say, "I really don't want to buy my next generation system from a vendor who is 8,000 miles away and doesn't speak my language, because I know for sure that it isn't going to work. On the other hand, etc."

Whether it is exploring for oil, refining it, or distributing and marketing oil, or whether it is making movies, or automobiles, the big moves in an industry always relate to the pattern of vertical integration. In 1920 MGM was 100 percent vertically integrated. It owned its own movie stars, its own lots and distribution system, and its own neighborhood exhibition houses. Today, it is vertically disintegrated. The oil companies in the late 1800s were disintegrated. By 1960 they were vertically integrated, and now they are disintegrating again. The auto industry was disintegrated, and then it was integrated, and now it is disintegrating again. These are always the big patterns that alter the structure of industries.

In 1985 a survey was conducted of twenty-nine Japanese chip makers, of whom twenty responded. There was one series of critical questions that bear on our current topic. That question asked the Japanese, "Where do you purchase your semiconductor tooling?" The answer was, "Today, we buy 23 percent of our tools from importers"—that means from U.S. tool makers. In the future, these Japanese chip makers expect to buy no tooling from abroad. The respondents also indicated that in 1985 they bought 57 percent of their tooling from domestic suppliers who were not members of their financial group. They expected that to rise to 89 percent. They also indicated that in 1985, they bought 19 percent of their tooling from a domestic supplier who was in "their group." They expected that to decrease to 6 percent.

What do the Japanese respondents mean by "in our group" or "not in our group"? "In our group" is the typical quasi-vertical integration that one sees in industry after industry in Japan and that is greatly underused in the United States. This is a financial form that represents a combination of the independence of competition and the capacity for teamwork. Financial bonds are the links that enforce a vertically integrated system. For example, Fujitsu owns 21.5 percent of the equity shares of one of its key vendors Advantest; Hitachi owns 21 percent of Kokusai Electric; and NEC owns 51 percent of Ando Electric.

In the semiconductor industry for the chip maker to have a good relationship with the tool maker, he has to be able to disclose proprietary information. Likewise the tool maker has to be able to disclose proprietary information to the chip maker. In addition, the chip maker may want the tool maker to make some specialized capital investments that suit only the chip maker's needs and nobody else's. On the other hand, the tool maker might like the chip maker to make some scheduling changes that suit only the tool maker's needs and nobody else's. If each firm will accommodate itself to the other's constraints, they are both better off. However, neither will do this if each thinks that the other may walk away on a whim. With this kind of quasi-ownership, you won't have control, but you will have influence. You have an exchange of hostages, if you will, a collateral bonding, which gives each the confidence that the other is sufficiently tightly linked. The relationship is loose enough so that if it is a fundamentally bad relationship, it can and will be broken. It is an intermediate form of vertical integration. These two parties work together and develop some new products, which one of them sells not only to his "partner" but also to his "partner's" competitors.

For example, an integrated circuit house may work closely with a tool maker to jointly develop some "bell or whistle" that goes on a tool. The logic is that the tool maker is typically a little company while the chip maker is a big company, and if it does not let the tool maker sell his inventory to everybody who will buy it, then the chip maker is going to end up having to fund all of the tool maker's future product developments, an option that is not all that attractive to the chip maker. Also, if the tool maker becomes weak, the chip maker will become weak. So what the chip maker really needs is an intermediate relationship in order to get about a two-year head start on the competition, as a result of his ties to the tool maker. This is very different logic from what we typically see in the United States. Usually in the United States a tool maker is required to sign in blood that he will never reveal, sell, or make available to anybody else any of the defined intellectual property, period.

COOPERATIVE R&D VENTURES: THE UNITED STATES AND JAPAN

In 1961 the Japanese government passed a special law that encouraged the formation of joint R&D ventures—horizontal ventures, I call them—among companies that normally are competitors. These ventures are meant to recognize that there are, in addition to public properties and private properties, a large class of what I would call generic properties—that is, property that has "leaky" characteristics. It is a property that does have private monopoly value to its inventor but just for a very short time. There is no practical way to prevent others from using it for free. Therefore, if you rely on private investment to create it, you will create it at a relatively low rate.

If in fact such R&D with generic, leaky properties is of high value to industry as a whole and if one country's industry does not create very much of it, but another country's industry does, then the country without it will lose in the competitive marketplace. When you have a leaky intellectual property, you should not seek to have it funded exclusively through private means or it will be underinvested. People will not pay to invent something that their competitor can use for free. On the other hand, if it is a generic property, you should never permit it to be funded entirely through public funds. If I can be compensated 100 percent for every nickel, direct or indirect, for a research project I undertake and then make $1 of private profit out of it, I have an incentive to use $1 million of public funds to create one dollar of private profit because it is net. If that happens, we will, of course, have an overrequest for those public funds for the development of such property and we will then have to institute an allocation mechanism, which will inevitably become a centralized decisionmaking process in Washington. I think that we all know intuitively, if not explicitly, that in such situations decentralized decisonmaking is far superior and that any kind of a central planning process will fail. Therefore, to achieve a socially optimal level, generic property must be funded and organized in a way that provides a mixture of public and private funds.

Japanese joint R&D projects have typically lasted four or five years and occasionally as long as ten years as in the case of the fifth generation computer project. (However they are legally held open beyond the typical four- or five-year limit in order to have a device through which to assign patent rights, royalty payments, and so on.) Past examples include the automatic graphic processing project, a large-scale computer that produced the M-series mainframe, the supercomputer, and the pattern information processing project. Each of these projects brought together major horizontal competitors such as NEC, Oki, Hitachi, Fujitsu, and NTT to produce generic properties that no one of them had a sufficient incentive to produce alone at an optimal level. These early projects produced frequent arguments among the principals, and they are considered not to have been great successes. But there was the perceived necessity. The companies felt that they had to work together, and in these early cases they were learning how to do it so they could do better the next time.

Unlike the U.S. projects that are almost always open-ended, lasting potentially forever, the Japanese joint ventures are all required by law to be close-ended. One consequence is that Japanese R&D projects tend to have highly focused goals, which can be stated very simply. The goal of the fifth generation computer project for example, is to come up with a computer that can recognize multiple human speakers, each speaking at up to three times the normal rate of speech, with an initial vocabulary of 50,000 words, complete syntactical rules and an error rate of less than 2 percent. You can understand and identify with that. You will know if you have it or not. It focuses the efforts of many different kinds of people in a very crisp way. Some other joint Japanese R&D ventures have focused on the following areas: business management software, office management software, design calculation software, operations research software, automatic control software, automatic measurement, VLSI, medical equipment labs, laser production systems, computer fundamental technology.

Following is a more detailed discussion of one of these Japanese joint R&D ventures that concerned high-speed scientific and technical computers and automatic sewing systems. Currently the United States and Japan are the world's low-cost producers in the high-end textile business because of tremendous automation. The looms in these textile mills are now operated with focused air jets. A puff of air catches the end of the thread and fires it across the loom up to 300 times per minute. It works with cotton as well as artificial fibers. However, the making of garments is still 70 percent by hand because of materials handling. Nobody has been able to automate that. With a wage rate of twenty-six cents per day in the People's Republic of China and a wage of $6.50 per hour in North Carolina, the U.S. garment making has gone overseas. The garment maker overseas sources his textiles locally. He is not going to ship his textiles from North Carolina to Singapore, get it into Shanghai, and then bring the finished goods back to sell in Texas. What happens is that even though we and the Japanese have the world's low-cost textile industries, both countries are losing this and the garment industry because, just like the semiconductor industry, the pieces are linked together into a vertical stream.

In response to this threat the Japanese have a fifty-company consortium made up of companies that make textiles, artificial fibers, garments, robots, sewing machines, lasers, computers, machine tools, and software. These companies are working together to try to develop the world's first "lights out" garment manufacturing plant. If you have ever seen the Fujitsu manufacturing plant in Japan, you know that it is truly a "lights out" factory. It is a robotic factory that builds robots with a first shift of three people, a second shift of about three people, and a third shift of zero people. They literally turn off the lights to save electricity costs, and the plant runs in the dark. It is real automation. That is what they have in mind for the garment industry. This consortium is funded with something like $250 million.

We have a similar effort in the United States, which has brought together about fifty U.S. textile and garment companies with a couple of unions and has total funding of about $1 million per year. We are not playing this game as seriously as the Japanese at the moment.

CONCLUSION

Joint R&D projects are organizations that the Japanese have had to learn how to govern, to structure, and to finance. They have had to learn by trial and error. It is clear that for the United States the time for debate has passed. The Microelectronics Computer and Technology Corporation (MCC) has set a pattern from which everyone else in the United States has learned. Its example has given inspiration to the entire country. The United States has thirty-five joint R&D ventures organized under the 1984 law (National Cooperative Research Act) inspired by MCC. This is a major new development. I think we are going to see several hundreds of these kinds of ventures coming forth. For example, Sematech, an industrywide consortium to develop state-of-the-art manufacturing technology, is moving forward in its attempt to develop the most advanced semiconductor manufacturing technology in the world. There is a large class of new "leaky" generic intellectual property that is critical to the development of the technologies that lie in our future. I expect that we are going see, in our lifetime, a whole new structural form emerge. It is going to be a very exciting period.

Chapter 15

TECHNOPOLEIS
Themes and Conclusions
David V. Gibson and James W. Dearing

The chapters in this volume provide a series of case studies and analyses of technopoleis at different stages of maturity in different national locations. Are these technopoleis cities of light and wisdom where scientific research and invention combine to produce new theoretical knowledge and innovation through practical applications? Can such regions approach the utopian vision of Companella's *Civitas Solis* (1623)? Or are such cities of technology more likely to be paralyzed by quality-of-life issues and to develop as two-tiered societies populated by elite Ph.D.s working in prestigious research institutions and unskilled workers, employed in low value-added, repetitive jobs and services? Taking the more optimistic view, Abetti and his co-authors emphasize the continuity between the mythical city of light of the Renaissance and illuminism and the modern technopoleis of the latter part of the twentieth century.

Technopoleis, and the innovative relationships nurtured within them, are considered a vital, generative component of a country's or a region's competitiveness in the global economy. Also, stressed in the nurturing and growth of technopoleis is the importance of university/industry ties. Although these are dominant themes of this book, they are not universally accepted.

The question of control over knowledge distribution is important not only to university governance but also to the nature of evolving technopoleis. Some policymakers argue that by their very purpose, private business and universities are in fundamental contradiction regarding knowledge. In order to maximize profits, businesses must conceal research results. Universities, to maintain their legitimacy and prestige, must freely disseminate state-of-the-art research findings. The interests of government in knowledge distribution are mixed, depending on the type of research being funded.

Ouchi emphasizes that the structure of intellectual property rights that reside in any society is a tremendously powerful force and that change in the structure of these rights is a fundamental fact of socioeconomic evolution. He gives persuasive examples and argues for the benefit of cooperative R&D ventures that have a "generic, leaky" character so that society and private investors both may benefit. He commends the United States for recognizing the value of using taxes and private funds to assist universities in creating new intellectual property even though these products often benefit some groups more than others. For example, Ouchi credits U.S. universities with putting the "gold" in the hills of Silicon Valley. It was this gold, or intellectual property, that was later mined by entrepreneurs and made possible one of the most famous, successful, and largest of the world's technopoleis.

Segal comments on the benefits of the way that Cambridge University operates with a benign, supportive, and noninterventionist policy regarding its faculty and "their" intellectual property. In the end, the technopolis at Cambridge is seen to benefit tremendously from this university's culture of excellence and openness. Even in the highly formalized education system of Japan, Tatsuno cites Yamaguchi University's technology center where professors and graduate students are free to conduct their own creative research outside university constraints. Beijing's Institute of Information and Control (BIIC) is experiencing favorable results with increased individual responsibility and accountability on research projects that are specifically designed for end-users. BIIC is also benefiting from encouraging a policy of openness through national and international exchanges, conferences, and educational programs.

Onda's comment that the development of technopoleis is constrained by a shortage of "brains" speaks directly to the centrality and scarcity of knowledge. High technology firms locate near research universities for very good reasons: access to knowledge and professionally trained personnel. The role of the research university in an information society has been compared with that of a factory in an industrial society. But the capacity to increase knowledge output is not analogous with the capacity to increase, say, textile output. Botkin stresses that money and brains drive economies and such resources can only be accumulated slowly and painstakingly. Furthermore, increasing the amount of basic research at universities has political, social, and economic ramifications. As the chapter by Smilor and others confirms, a commitment by state government to increased funding for higher education builds momentum that can have a multiplier effect on a state's economy. Conversely, when the state reduces its level of support for the university, "the wheel of technopolis progress slows."

How important is long-range planning to the nurturing and maintenance of technopoleis? And if it does occur, how effective is it? As advocated by Ryans and Shanklin there is evidence of careful, long-term planning being used in the development and management of technopoleis in the United States, such as the Arizona State University Research Park and the Rensselaer Polytechnic Institute, and elsewhere in the world (for example, Sophia-Antipolis, Tsukuba Science City, the Osaka Technoport Project, the Naniwa Necklace Project, and Kansai Science City).

Since passage of the technopolis law in 1983 and the enactment of twenty-year development plans, Japan is perhaps the most ambitious nation in actively planning for "cities of light and wisdom." However, based on the range of examples discussed in this text, the benefits of such long-term planning are often mixed. For example, in one of the most planned technopoleis, Tsukuba City, there is evidence of unintended, dehumanizing consequences as manifested in Tsukuba syndrome, where the city's inhabitants actually develop a skin rash. Furthermore, the suicide rate in Tsukuba City has been higher than anywhere else in Japan and is purportedly a result of a sterile environment that is lacking the Japanese amenities offered by older, less planned communities.

On the other hand, there stands Laffitte's example of the bistros and gardens that were planned as integral parts of Sophia-Antipolis to make this technopolis a place where high-quality community life satisfies a range of human and scientific needs. And there are the Japanese examples of concern for open space and greenery as paramount in the construction of technopoleis of the future. These planned areas of beauty are considered important facilitators of informal face-to-face communication. Larsen and Rogers consider such informal intra- and interorganizational communication networks as crucial to the success of Silicon Valley. As a more formal method of information exchange Tatsuno, in describing the building of a Japanese technostate, emphasizes regional data bases and on-line information centers to facilitate information exchange within and between technopoleis. Indeed, the importance of social and interdisciplinary "networking" within and between the academic, research, and business communities is a common theme of this volume.

There is also evidence in the United States and other countries that technopoleis are born and mature where there is a noticeable lack of planning, such as Silicon Valley, Cambridge, and Route 128. However, once again it is clear that these areas also develop a range of problems engendered by their very success, such as traffic congestion, pollution, a rising cost of living that precipitates a two-tiered society, and a cyclical economy that is closely tied to the fortunes of particular high tech industries. So, while the spontaneous development of cities of technology may solve, in a serendipitous way, the dehumanizing problems that some highly planned technopoleis face, the community is then left with choices about long-term growth and an affordable quality of life that necessitates coordinating community, business, university, and political coalitions after the fact.

Recently, the national press has publicized the connection between environmental pollution and high technology manufacturing. Toxic waste, lethal gases, and other contaminants pose real and potential problems to many communities. And as Rogers and Larsen, and Botkin note, newer technopoleis do not seem to be learning from the quality-of-life lessons of Silicon Valley and Route 128. According to reports on the more mature technopoleis, emulators would benefit most from learning from the historical mistakes of such areas rather than trying to replicate their peculiar successes.

In short, an important question is to what degree can or should the culture and infrastructure of a technopolis be planned and to what degree must it arise spontaneously? The desired product is knowledge. Creative thinking is stimulated by attractive cultural, social, and environmental stimuli. Lacking such stimuli, can technopoleis be successful?

Although quality-of-life issues provide an important theme that crosses all the chapters of this book, it is a theme that is not easily defined. Most of the technopoleis discussed here, at least by their geographical locations are cities of the sun. They have risen in industrially virgin areas where entrepreneurial ventures tend to be visible and easily nurtured. But this is not always the case. As Botkin notes, each successive industry in the Boston area has been built on the remains of proceeding industries. Rensselaer Technology Park, also located in the northern United States, is being built in an industrially mature area that is simultaneously being abandoned by large manufacturing companies.

Are the current forms of high technology industry ephemeral? The functions of firms remaining in Silicon Valley or Route 128 are very different when compared to the activities that took place in those areas fifteen years ago. However, while activity has changed as a result of area and industry maturity, vital signs of entrepreneurial life still exist. The number of new firms locating in Route 128 and Silicon Valley is still high.

Botkin states that over the long run Route 128 has been home to economic cycles of boom and bust, each ridden by a different industry (the current industry being high tech). Only universities and banks ("brains and money") have remained. But as the experiences of other technopoleis (such as the Austin/San Antonio Corridor) suggest, without some form of economic base to stimulate growth and employment, tax income shrinks, governments cut back on education funding, and bank failures increase. It is also interesting to note that it is during (if not before) such times of economic "bust" that government policies concerning educational and other incentives may be most crucial in developing the resources needed to spur the growth of technopoleis.

The questionable future of the Cambridge phenomenon and the emerging Austin/San Antonio technopolis highlights the importance of maintaining a university's international reputation for research and teaching excellence. Mark stresses the crucial resource of a few very bright people who spark an aura of intellectual excitement and who ask the visionary questions that if answered dramatically change the way that we live. But as Segal points out, the university has limited power to redress adverse state and/or national financial stringencies that make it difficult to be able to afford to recruit such "national resources" and to sustain their work.

The case of the Rensselaer Polytechnic Institute (RPI) provides a vivid example of a technopolis's being born during a period of economic stagflation. In this area of declining industrial strength and rising unemployment, RPI through the incubator program and the Rensselaer Technology Park has added to its roles of education, and research and technology transfer, by stressing a new active role of promoter of technological entrepreneurship and regional economic development. Unlike many other technopoleis that have been developed around one technology, the RPI model is diversified through strong ties with the vast array of resources available at the university. At a broader policy level, to counteract the recessionary effects of the "yen shock," the Japanese government recently announced an elaborate set of fiscal policies as a pump-priming package designed to facilitate the regionalization of high tech research. Tatsuno sees this as a major policy shift that will have a strong impact on the global economy.

The conceptual framework of a Technopoleis Wheel focuses attention on crucial sectors in the process of technology commercialization and economic development in the Austin/San Antonio Corridor. The identified segments are (1) the research university, (2) large corporations, (3) emerging companies, (4) the federal government, (5) state government, (6) local government, and (7) support groups. Depending on how these seven segments interact, each may contribute to or constrain progress in a technopolis.

Numerous authors have noted that short-sighted, fragmented policymaking among the government, business, and university components of a technopolis is more common than is a harmonious working relationship. The Austin/San Antonio Corridor's experience in "winning" MCC indicates that such cooperation across the Technopolis Wheel is possible. But the more recent experiences of this area also raise serious questions as to whether such synergy can be maintained over the long term.

It is unclear how precisely the Technopolis Wheel framework maps to other regions in the United States and which segments pertain to the development and management of technopoleis in other regions of the world. Although it is striking how much the case studies from different U.S. sites and from Japan, China, England, and France would seem to suggest that the same dynamics apply. Most important, two fundamental facts seem to be especially robust. One is the importance of high-quality research universities and the crucial underpinning resource of a few very bright people who are engaged in basic research. There also is the key role that the university plays in providing the talent and professionally competent and managerially adept people to combine scientific research and invention with the practical applications of technology. And the university is an important source of liberal arts that underpin the quality-of-life factors necessary to sustain technopoleis. Although a preferred geographical climate is not consistent across all the examples of technopoleis, a high quality of intellectual and cultural stimulation is.

The second consistent theme is the importance of a network of influencers or "executive champions" from the different segments of the business, academic, and government communities. These influencers provide the inspiration and vision necessary for nurturing and maintaining a technopolis. At the core of the Technopoleis Wheel, these influencers link the seven sectors. It is the task of these influencers to make the Technopoleis Wheel spin at the right speed and direction toward balanced growth and development. Ryans and Shanklin stress that it is crucial for such influencers to be able to separate unrealistic individual and community dreams and aspirations from the real-world challenges of fostering and managing successful technopoleis and a desirable, affordable quality of life.

Agreeable compromise across the various sectors that affect the nurturing and growth of a technopolis is clearly not easily achieved and sustained. Solutions to problems most likely vary with the peculiarities of different regions. An intriguing question is whether the reality of "cities of light and wisdom" is more within the grasp of today's influencers than yesterday's visionaries who identified and began the quest.

INDEX

ABOUT THE SPONSORS

THE UNIVERSITY OF TEXAS AT AUSTIN

The IC^2 Institute is a major research center for the study of innovation, creativity, and capital (hence, IC^2). The institute studies and analyzes information about the enterprise system through an integrated program of research, conferences, and publications.

The key areas of research and study concentration of IC^2 include the management of technology; creative and innovative management; measuring the state of society; dynamic business development and entrepreneurship; econometrics, economic analysis, and management sciences; and the evaluation of attitudes, opinions, and concerns on key issues.

The institute generates a strong interaction between scholarly developments and real-world issues by conducting national and international conferences, developing initiatives for private- and public-sector consideration, assisting in the establishment of professional organizations and other research institutes and centers, and maintaining collaborative efforts with universities, communities, states, and government agencies. IC^2 research is published through monographs, policy papers, technical working papers, research articles, and four major series of books.

The College and Graduate School of Business prepares outstanding students for a variety of careers and early assumption of management responsibilities in a rapidly changing environment. A continually evolving curriculum ensures dynamic programs that meet the needs of future business leaders. A broad range of elective opportunities enables the student to take a generalist approach to advanced study, to concentrate in a traditional or emerging field of business study, or to broaden perspectives through supporting coursework in nonbusiness disciplines.

The College of Business Administration and the Graduate School of Business were created in 1922. The college has five departments: accounting, finance, management science and information systems management, and marketing administration. Approximately 8,640 undergraduate students, 1,030 M.B.A. students, 110 M.P.A. students, and 235 Ph.D. candidates are currently enrolled in the College and Graduate School of Business.

The Department of Management Science and Information Systems (MSIS) in the College and Graduate School of Business focuses on management science, information systems, and statistics. Its faculty has a record of research that has been accorded national and international recognition in the form of numerous awards, citations, and other honors. Based in applications that are relevant to managerial use, this research has produced new concepts and new approaches to problems with results that have also affected scientific research at basic levels in many different disciplines. New models and new methods of analysis and implementation developed by members of the management science faculty have been used as guides and incorporated in computer codes that are widely recognized and extensively used in both managerial and research activities in many parts of the world.

The faculty in information systems have developed new concepts and approaches that relate behavioral science and management organization and design approaches to the development and use of information and telecommunication systems. Statistics faculty are directing their research toward developing improved methods for identifying and structuring problems in the context of managerial applications and uses as well as producing improved methods for validating and interpreting inferences that may be made from data-based models and methods of analysis.

The applications through which these new developments are occurring are all related to computer uses and developments. Many results of this research are extending computer capabilities as well as computer uses in both management and science. Research by faculty on extensions of telecommunication capabilities include new uses of computerized methods for effecting transactions between business entities as well as controlling and directing activities within organizations.

THE UNIVERSITY OF SOUTHERN CALIFORNIA

The Annenberg School of Communications at the University of Southern California, Los Angeles, is a graduate-level training and research center, with special expertise on the diffusion of new communication technologies and on their social impacts. The school offers an M.A. degree in communication management and a Ph.D. degree in communication research and theory. While the Annenberg School is ideally sited in Los Angeles, the capital of the mass media industries, it also conducts research on a worldwide basis. Several of the school's faculty have a special interest in the "information society" that the United States and many European nations and Japan are becoming, and in the role of high technology industry in such a society.

RGK FOUNDATION

The RGK Foundation in Austin, Texas, was established in 1966 to provide support for medical and educational research. Major emphasis has been placed on the research of connective tissue diseases, particularly scleroderma. The foundation also supports workshops and conferences at educational institutions through which the role of business in American society is examined.

The RGK Foundation Building has a research library and provides research space for scholars in residence. The building's extensive conference facilities have been used to conduct national and international conferences. Conferences at the RGK Foundation are designed not only to enhance information exchange on particular topics, but also to maintain an interlinkage among business, academia, community, and government.

ABOUT THE EDITORS

Raymond Smilor is executive director of the IC^2 Institute at The University of Texas at Austin and serves there as a member of the faculty in the Department of Marketing in the College of Business Administration. He holds the Judson Neff Centennial Fellowship in the IC^2 Institute. Dr. Smilor earned his Ph.D. in U.S. history at The University of Texas at Austin.

He has served as a research fellow for a National Science Foundation international exchange program on computers and management between the United States and the Soviet Union and has been a leading participant in the planning and organization of many regional, national, and international conferences, symposia, and workshops. He is a consultant to business and government. He lectures internationally and speaks extensively to business, professional, and academic groups. He is also involved in several civic organizations and appears in *Who's Who in the South and Southwest.*

Dr. Smilor's publications have covered a wide variety of interdisciplinary subjects. His research areas include science and technology transfer, entrepreneurship, marketing strategies for high technology products, and creative and innovative management techniques. His works have been translated into Japanese, French, and Russian.

He is co-editor of five books: *Corporate Creativity: Robust Companies and the Entrepreneurial Spirit*; *Improving U.S. Energy Security*; *Managing Take-Off in Fast Growth Companies*; *The Art and Science of Entrepreneurship*; *Creating the Technopolis: Linking Technology Commercialization and*

Economic Development. He is co-author of two books: *Financing and Managing Fast-Growth Companies: The Venture Capital Process* and *The New Business Incubator: Linking Talent, Technology, Capital and Know-How.*

George Kozmetsky is executive associate for economic affairs in The University of Texas System. In addition, he also serves as director of the IC^2 Institute, professor in the management and computer sciences departments, and the J. Marion West chair professor at The University of Texas at Austin. He also is professor in the Department of Medicine of The University of Texas Health Science Center at San Antonio. Dr. Kozmetsky served from 1966–1982 as dean of the College and Graduate School of Business at The University of Texas at Austin. Dr. Kozmetsky received his bachelor of arts degree from the University of Washington in 1938 and the master of business administration degree in 1947 and the doctor of commercial science degree in 1957 from Harvard University.

His business acumen spans service, manufacturing, and technology-based industries. He is the cofounder, a director, and former executive vice president of Teledyne, Inc. He serves on the board of several other companies. He is an acknowledged expert in high technology and venture capital.

Dr. Kozmetsky is a fellow of the American Association for the Advancement of Science. He is a charter member and served as president of the Institute of Management Sciences (TIMS). He also is a chancellor of the American Society for Macro-Engineering and president of the Large Scale Programs Institute. He serves as a special reviewer for the National Science Foundation and is a member of the American Institute of Certified Public Accountants and the British Interplanetary Society.

Dr. Kozmetsky has served both state and federal governments as an advisor, commissioner, and panel member of various task forces, commissions, and policy boards. He regularly provides special testimony on business and technology issues to state and federal legislators.

He writes extensively. His articles and papers have appeared in major professional journals, magazines, and newspapers. His two most recent books are *Transformational Management* and *Financing and Managing Fast Growth Companies: The Venture Capital Process.*

David V. Gibson is an assistant professor in the Department of Management Science and Information Systems in the College and Graduate School of Business at The University of Texas at Austin. He received his B.A. degree from Temple University, an M.A. from the Pennsylvania State University, and an M.A. from Stanford University. In 1983 he earned a Ph.D. in sociology from Stanford after completing studies in the areas of organizational behavior and communication theory.

Dr. Gibson is co-organizer of the Annual Texas Conference on Organizations, which is sponsored by the College and Graduate School of Business at The University of Texas at Austin and includes organization scholars and graduate students from universities throughout Texas. Professor Gibson teaches undergraduate and graduate courses on communication behavior in organizations, international business communication, the management of information systems, and research methods in information systems. He belongs to the following professional associations: the Academy of Management, the American Sociological Association, the International Communication Association, and the World Future Society.

Dr. Gibson's research and publications focus on the management of information systems, cross-cultural communication and management, and the management and diffusion of innovation. He is currently working on a book with Professor Everett M. Rogers on how the Microelectronics and Computer Technology Corporation (MCC) decided to locate in Austin in 1983 and the state and national social, economic, and political effects of that decision.

ABOUT THE CONTRIBUTORS

Pier A. Abetti is a professor in management of technology and entrepreneurship in the School of Management at Rensselaer Polytechnic Institute in Troy, New York.

James W. Botkin is a partner and cofounder of Technology Resources Group of Cambridge, Massachusetts.

James Dearing is a doctoral student at the Annenberg School of Communications of the University of Southern California at Los Angeles, California.

Hiroshi Hiraoka is a city planner in the Department of the Comprehensive Planning Bureau, City of Osaka, Japan.

Pierre Laffitte is a senator of France as well as president and founder of Sophia-Antipolis.

Judith K. Larsen is president of Cognos Associates, Los Altos, California.

Christopher LeMaistre is director for the Center for Industrial Innovation at Rensselaer Polytechnic Institute, Troy, New York.

Hans Mark is chancellor of The University of Texas System, Austin, Texas.

Keisuke Morita is director of the City Planning Department of the Comprehensive Planning Bureau, Osaka, Japan.

Masahiko Onda is the principal research officer of the Mechanical Engineering Laboratory, Agency of Industrial Science and Technology, Tsukuba, Japan.

William G. Ouchi is a professor of management at the University of California, Los Angeles, California.

Everett M. Rogers is the Walter Annenberg Professor at the Annenberg School of Communications at the University of Southern California, Los Angeles, California.

John K. Ryans, Jr. is a professor of marketing and international business at Kent State University, Kent, Ohio.

Nick Segal is a partner in Segal Quince Wicksteed, economic and management consultants, in Cambridge, England.

William L. Shanklin is a professor of marketing and international business at Kent State University, Kent, Ohio.

Song Yuhe is assistant professor in applied mathematics at the Beijing Institute of Information and Control, Beijing, China.

Sheridan Tatsuno is a senior industry analyst for Dataquest's Japanese Semiconductor Industry Service in San Jose, California.

Michael Wacholder is director of Rensselaer Technology Park at Rensselaer Polytechnic Institute, Troy, New York.

Rolf T. Wigand is a professor in the School of Public Affairs at Arizona State University, Tempe, Arizona.

Yu Jingyuan is deputy director and professor in mathematics and control at the Beijing Institute of Information and Control, Beijing, China.

Zhou Zheng is an associate professor at the Beijing Institute of Information and Control, Beijing, China.